Lab Manual for
Connecting Chemistry to the Tribal Community
Two Semesters of Chemistry Experiments and Teachings

by Mark Griep, Bev DeVore-Wedding, Janyce Woodard, and Hank Miller

Keeper's Cottage Press, 2018

NEBRASKA INDIAN
COMMUNITY COLLEGE

LITTLE PRIEST TRIBAL COLLEGE
"BE STRONG AND EDUCATE MY CHILDREN"

Keeper's Cottage Press

About the Electronic Version

Chemistry instructors at tribal colleges and universities are free to use this electronic version and to distribute copies of it to the students participating in the laboratory portion of their courses.
All other chemistry instructors should only use this electronic version to review its content; they do not have permission to use the experiments in their classrooms.

About the Print Version

A print version of this lab manual is sold through Amazon.com and other venues.
Everyone can purchase the print version for use by students in their classrooms, including tribal colleges and universities who don't want the expense of printing copies of the electronic version.
The royalties from all purchases of the print version will be split between Little Priest Tribal College and Nebraska Indian Community College to support student learning activities.

About the associated "Instructor & TA Manual"

An associated manual for Instructors and Teaching Assistants includes information about setting up the experiments and preparing the solutions. It also contains example answers to many of the questions. A copy of the "Instructor & TA Manual" can be obtained by sending an email to Dr. Mark Griep at mgriep1@unl.edu.
The Instructor & TA Manual will only be sent to instructors who send the following information:
your name, your institution's name, a statement that you are an instructor at that institution, and the name of chemistry course and/or how you intend to use the lab and instructor manuals.

ISBN-13: 978-1721129065

ISBN-10: 1721129065

Table of Contents

General Course Instructions

HOW TO USE THIS LAB MANUAL

The labs in this manual were created for a two-semester General, Organic, and Biochemistry course sequence at Nebraska's two tribal colleges. While the authors see chemistry everywhere, we hope that through these series of experiments, the background and community connections, that the students not only see the chemistry every day, but also find enjoyment in doing chemistry. Whether you continue in chemistry or science, it is important that you understand not only how chemistry affects your daily lives, but how it can enhance the quality of your lives. The labs are adaptable and can be used by other tribal colleges and community colleges. The labs can be performed by students alone or in pairs and will require about 2.5 hours to complete if the reagents and materials are ready.

All labs have background information, community connections, the lab protocols and procedures, and suggestions for the lab report. The appendices include standard chemistry reference tables, a list of chemistry community connections, and lab report templates. The information regarding laboratory notebooks and scoring is provided as a guide that they instructor can adapt according to local needs.

The last lab in each semester is a creative project and presentation. Students should work on this throughout the semester. Therefore, a short proposal (roughly one paragraph) describing your plans for your project should be due at midterms. Your lab instructor should approve your project selection. Instructors may want to check in with the student more frequently to monitor progress as well.

We invite users to provide feedback on these labs and community connections. We also welcome the exchange of ideas about new labs or community connections.

For the Instructor

This laboratory manual was written to accompany a two-semester general, organic, and biochemistry course sequence. Several additional labs were added to emphasize concepts that are important foundational knowledge and to increase proficiency using standard laboratory equipment.

The importance of engaging students through hands-on laboratory experiments and connecting to their personal experiences cannot be overstated. Pre-lab discussion should emphasize safety, following protocols, introducing any new equipment and procedures, and connections to the students' personal experiences and community connections. Specific community connection information has been provided in the background of the lab, but each lab report provides students the opportunity to share their own personal connection. The accompanying instructors' manual provides information about equipment and materials, preparation of each lab with safety reminders and precautions, solution and sample preparations, and possible answers to the questions from the conclusion sections of the labs. In addition, discussion suggestions for community connections and sharing, additional references for background reading connecting to authentic application for the specific labs, and an equipment and materials inventory is included in the appendix.

The first eleven labs have specific student report form templates, which are found in the appendix. These have been provided for students who may not have had previous experience writing lab reports; as one progresses through the first semester, each lab requires the student to provide more of the lab report itself until finally they are writing the entire lab report without a specific lab report. A generic lab report template is included for subsequent lab reports (Appendix D).

This lab manual has been a collaborative effort involving several people and their own curriculum resources. Dr. Mark Griep provided the framework and many of the labs are from his own teaching resources. Janyce Woodard co-taught and aided in our revisions to accommodate community connections. Hank Miller supported our work and shared his copy of the RISE experiments from Turtle Mountain Tribal College from which we borrowed ideas and revised our own labs. LaVonne Snake assisted in tribal community connections and in the teaching of the labs, providing important feedback. Qudsia Hussaini also assisted in the teaching of the labs, providing important feedback.

For the students

Your own connections to the topics of the labs in this manual are important in connecting the science to the real-world, to understanding the importance of chemistry to our everyday lives, and to authenticate the laboratory experience. The community connection section of each lab only scratches the surface of the tribal community connections. The sharing of your own connections will enrich the authenticity of the chemistry as well as your instructor's and classmates' own knowledge.

Each investigation has stated objectives, which the student should keep in mind while performing the experiments, collecting and analyzing their data, and summarizing the lab in your report. In addition to the step-by-step lab protocols, safety precautions and awareness are emphasized in each lab and as a starting point for the manual itself. The labs have been written to avoid hazardous materials as much as possible and when they are used, to decrease the amount of chemical used. However, your safety really is based on you performing your labs safely as well as following the precautions and the directions of your instructor.

Lab report templates are available for the first eleven labs to provide support for those who have not written lab reports previously; a generic template is also provided to aid in your reporting your lab findings. Questions in the conclusions guide your analysis of the data you collected. Your data should be organized and for most labs data tables are provided. These tables may be copied for your lab reports. If students are directed to modify the procedures or substitute chemicals, these changes should be noted in their lab reports as well.

COMMUNITY CONNECTIONS

An advisory board of community representatives met and provided a list of community topics (Appendix B) for this project. These topics provided the foundation of chemistry community connections in the individual labs. The community connection, Life Flows (Appendix C), provides specific community connections for Nebraska Indian Community College and Little Priest Tribal College as well as basic water chemistry content (Teacher notes are in the Instructors' Manual). However, these topics are global and may be adapted to any community's unique and individual interests.

Why community connections? Connecting chemistry content to students' a prior knowledge, authentic phenomenon, and culturally-relevant topics increases the learning of new content. Providing laboratory experiences that tap into the students' own experiences enriches the learning experience. Some of the labs are easier to connect to the community topics than others. Included in each lab report is an opportunity for students to share their cultural and community experiences that connect to the lab. Both Instructors and students are encouraged to share their connections in class as well as with the authors of this manual. One goal is to share community connections beyond the tribes served by Nebraska Indian Community College and Little Priest Tribal College to increase the depth and breadth of tribal community connections with chemistry.

Evergreen State University has a repository of community connections, Enduring Legacies, Native Cases. These case studies (what we call community connections) are mono-topic, unlike the community connection accompanying this manual, and are available for use. Additional case studies for science instruction may be accessed at the University of Buffalo's National Center for Case Study Teaching in Science.

LABORATORY NOTEBOOK

Each student should have a three-ring binder with this lab manual inside. You are responsible for all work, even when working with a partner. During lab, remove just the lab instructions and recording sheet to give yourself room at your lab space, keep track of procedures, equipment, and materials, and record your observations, measurements, and other data. Your raw data will be used to complete the lab report that you turn in the following week. When working with a partner, one person may record all of the data, but before you leave the lab, all data needs to be shared so each person has their own set of data. While you may have partners for your lab experiments, your lab report should be your own. *Your answers may be like your lab partner's but should not be word for word the same.*

Every experiment has a data recording sheet and the first eleven labs have a clean template for your actual lab write-up (see Appendix D). As we progress through the course, you will be required to prepare more of the lab report until you only have questions and the lab template for your lab reports. Typed labs are preferred as they are easier to read and may be turned in electronically.

Lab Report Scoring Instructions

Every lab and report are important. *If you miss a lab and turn in a report using data gathered by the class, **you must reference** your source for data in the report.*

Good lab reports will include

1. your name, date, class & section, partner's name, experiment title;
2. the purpose or objectives of the lab;
3. materials and equipment with precautions; ***only add if modified from as given; otherwise say 'as written;'***
4. procedures followed ***only add if modified from as given; otherwise say 'as written;'***
5. results (data) in a table and/or graph format;
6. conclusions are a discussion that describes how the results are related to the objectives of the experiment. You should also include answers to the questions specific to the lab itself;
7. references for information you looked up elsewhere and for assistance you may have had from your partner, TA, or another person.

All labs are due two weeks after completion of data collection.

Lab Report Scoring Criteria

1. Points total possible, including points for pre-lab safety quiz;
2. Formatting: your name, your partner's name, experiment title, and date you collected data; the purpose or objectives of the lab; any changes in procedures noted; any changes in materials and/or equipment
3. Results: Observations and data recorded in table, correct units given, figures labeled, work for at least one of each type of calculation or referenced to website used for calculating; errors discussed, if any; answers to report questions, and 2-5 sentence summary of results.
4. Penalty: ***points*** of for every week report is late. Report will be due two weeks after data collection.

TRACKING YOUR LAB REPORT SCORES

This section is for keeping track of your grades in lab.

Lab Title	Points Earned	Points Possible

ABOUT THE PAINTING ON THE COVER

The painting on the cover was created expressly for use by tribal colleges to conceptualize how science courses should be taught. Its title is "The Sharing Cycle of Science Learning" and it was painted by Laurie Houseman Whitehawk, who is affiliated with the Santee Sioux and Winnebago tribes. Her paintings are in the collections of the Museum of Nebraska Art in Kearney and the Great Plains Art Gallery in Lincoln. Prints of her work are prominently displayed in many buildings in northeast Nebraska. The painting provides a visual description of the "Framing the Chemistry Curriculum" project being carried out by Little Priest Tribal College and Nebraska Indian Community College.

The painting is in the form of a medicine wheel with its four sectors, each signifying a different aspect of a cycle but also signifying unity at the center of the wheel. The "Framing the Chemistry Curriculum" project is a cycle with four parts, each of which needs to be shaped to create stronger ties between the communities and the science taught in their tribal colleges.

In the first sector (upper left), a government official provides a boon to a native woman in the shape of an Erlenmeyer flask and she gives him a gift in return. This sector represents the National Science Foundation grant that funds the project but also the role of the community in shaping it. There is an Advisory Board composed of community and tribal leaders that meets once a year to generate and manage a list of Community Topics. The Advisory Board also review the previous year's activities.

In the second sector (upper right), the Community Connections Group composed of faculty and students meets once a year to identify scientifically measurable parameters that are related to Community Topics. The goal is to find materials and measurements that are important to life in the community.

In the third sector (lower right), the two-semester Chemistry Course sequence is offered at both Little Priest Tribal College and Nebraska Indian Community College. The laboratory experiences are designed to analyze locally relevant materials in ways that are useful to the community.

In the fourth sector (lower left), the Faculty Workshop composed of instructors and teaching assistants meets to share what they learned about teaching chemistry to tribal college students. It also works to strengthen the ties between the course content and the communities being served.

FIRST SEMESTER LABS

1. SAFETY, EQUIPMENT, AND MEASUREMENT

Part A: Safety & Equipment

I. **Objectives**
   1. To understand safety protocols and safety features of your classroom
   2. To understand safe laboratory behavior

II. **Facts to know**

Instructors are responsible for classroom and laboratory safety which is why students are required to follow all safety rules and instructions. Personal protective equipment, safety glasses or goggles, chemical-resistant gloves and aprons or lab coats, are for student protection and are required for participation in labs.

In any laboratory where chemicals are used, there will be hazards. Whenever possible, the least hazardous chemical or material has been incorporated into the labs. Each lab has a safety section with appropriate information specifically for the lab as well as in general. Your instructor will remind you of the necessary safety procedures however, by reading each lab prior to class, will increase your safety.

The American Chemical Society provides guidelines you may peruse to enhance your safety knowledge, such as Safety in Academic Chemistry Laboratories (https://www.acs.org/content/dam/acsorg/about/governance/committees/chemicalsafety/publications/safety-in-academic-chemistry-laboratories-students.pdf). If safety is of interest to you, there are even careers in safety management and enforcement in academia and industry. For further research you can contact the Occupational Safety and Health Agency (https://www.osha.gov/), Mine Safety and Health Agency (https://www.msha.gov/), or search for environmental health sciences.

III. **Community Connections**

Safety within your community may encompass personal safety through community-wide safety. Personal safety includes a wide variety of activities such as following directions using household products, maintaining a healthy lifestyle, operating mechanical equipment correctly, walking facing traffic, and staying aware of one's surroundings. Community safety includes weather alerts such as public warning systems for disasters, natural or man-made, neighborhood watch programs, ride-sharing programs that provide transportation for those without vehicles or towns without public transportation, and community centers.

Review the community topics (Appendix B) to identify possible community connections.

IV.   **Safety Concerns**
1.   Wear eye protection.
2.   Locate safety equipment in classroom.

V.   **Materials**

Equipment   **Samples and Reagents**
Lab Room   Safety information
Safety quiz

VI.   **Procedures**
1.   Review safety procedures for your school and class.
2.   Complete the report form's drawing of lab room & equipment page (Appendix C)
3.   Read the section on Specific Safety Equipment below
4.   Take the Safety Quiz

Specific safety protocols and equipment
1.   <u>Fume Hoods</u> - chemical fume hoods are used when you are working with chemicals with strong odors, are toxic, or flammable, by themselves or during a chemical reaction.
2.   <u>Sinks</u> – do not pour anything down the sink without first checking your lab instructions and or instructor; for the most part, we will use chemicals that are not toxic and can be disposed in the sink. Do not throw paper or solid materials into the sink; keep it clean!
3.   <u>Broken Glass</u> – do not throw into trash; if on floor, sweep it up, and place in special receptacle labeled "Broken Glass." This keeps you and the person emptying the trash safe from cuts! If you cut yourself, seek help immediately from your TA or Instructor.
4.   <u>Eye Wash/Shower</u> – for washing chemicals you may spill on you or splash into your eyes; flush your eyes for a minimum of 15 minutes; if chemicals spill on your clothes is severe, remove clothes and shower.
5.   <u>Spilled solutions</u> – IF you spill a strong acid or base, we need to neutralize it before wiping it up; for a strong acid, add enough baking soda (a weak base) to soak up all of the acid solution; for a strong base, add enough vinegar (a weak acid) to lower the pH (which means you should test the pH before wiping it up). Any other spilled solutions should be reported to your TA immediately for clean-up instructions to prevent accidents.

6. Use of Chemicals (Reagents) – always read labels of solutions before you actually use any reagent; if you are given the original container of a reagent, put a smaller amount into a beaker for your use during the lab; this makes pouring easier and reduces contamination of original reagent; if pipetting solutions, do not touch the tip of the pipettes to the counter surfaces, glassware, etc. Rinse graduated cylinders, beakers, and flasks well before using it with a new reagent. Do not return excess chemicals to original container. Dispose according to your lab, TA's or instructor's instructions.

7. Glassware – before you begin your lab, check your glassware for cracks, chips, and cleanliness. Too much detergent in washing glassware is a common contaminant in experiments but be sure to clean your glassware after use carefully and with detergent. Rinse soap off with plain tap water and a final rinse with distilled water. Place glassware upside down to dry in the rack provided.

8. Tap water versus Distilled water – unless specifically stated, use distilled water for all labs.

9. Goggles –eye protection should be worn from the beginning of the lab until the end. If you are concerned about sanitation, you may purchase your own pair.

10. Gloves – protect your skin from chemicals; if you are allergic to latex, please tell your TA or Instructor and make a note of it on your safety quiz.

11. Lab coats –protect your clothing and skin; closed toed shoes and long pants/skirts should be worn during labs;

12. Safety Preparation – develop these habits of mind and action to stay safe:
    a. *Read your lab prior to class! Ask questions!*
    b. *Do not start a lab without TA/Instructor present.*
    c. *If you have an accident*, notify your TA/Instructor immediately.
    d. *First Aid Protocols:* Tell your TA/Instructor immediately.
    e. *Pull your hair back* and roll up your loose sleeves.
    f. *Wear* proper clothing including a lab coat, gloves and goggles.
    g. *Do not drink or eat* while working in the lab.
    h. *Never taste chemicals* or mouth pipette.
    i. *Do not smell flasks, test tubes, etc. directly*. Waft the vapor carefully toward yourself. Treat all gases and vapors as though they will knock you unconscious if you breathe them.
    j. *Point reaction tubes* and containers away from yourself and others.
    k. *Keep combustible materials* and chemicals away from open flames.
    l. *Treat all solutions* as though they are corrosive toxins.
    m. *Always add things to water*. Never add water to strong acids or bases because the heat generated may cause them to splatter.

n. *Hot glassware* looks just like cold glassware.

o. *Water looks the same whether something is dissolved* in it or not.

p. *Record all observations and data in your lab.* This includes weight measurements, temperatures, color changes, physical state changes, etc. Check to see that your data makes sense before you clean up!

13. <u>Proper Handling of Equipment and Materials</u>:

a. Test equipment is designed to provide years of dependable service. Following these suggestions will help increase equipment performance.

b. Carefully follow all instructions.

c. Do not handle tablets; dispense from cap to test tube.

d. Carefully wash and rinse all apparatus used.

e. Tighten reagent caps immediately after use. Do not interchange caps.

f. Avoid prolonged exposure to direct sunlight.

g. Avoid temperature extremes. Protect all test components from freezing.

h. Anticipate your requirements for replacement reagents.

i. Keep all reagent containers out of reach of young children.

j. All chemicals have a safety data sheet (SDS)that provides chemical, physical, and safety information. An example of a SDS will be provided by your instructor.

Part B. Measurement

I. Objectives

1. Familiarize yourself with equipment you will be using during this course.
2. Practice measuring volume using graduated cylinders and pipettes.
3. Evaluate the accuracy of your measurements.
4. To calculate the mean and standard deviation of multiple measurements of the same substance.
5. To learn about significant figures.

II. Facts to know

Measurements

Accuracy and precision is important in measurements in the lab. ***Accuracy*** is the ability to measure the correct value and ***precision*** is the ability to repeat that measurement. While laboratory containers have markings indicating volume, some are more accurate and precise than others.

Liquids in some glassware may adhere to the sides to form a curved surface, called a meniscus, making a measurement questionable. The accepted measurement protocol is to measure the bottom of the curve as shown in the figure to the right.
In this lab, students will practice measuring liquids while also identifying which graduated (measurement markings) container is most accurate. Students will also determine precision by averaging

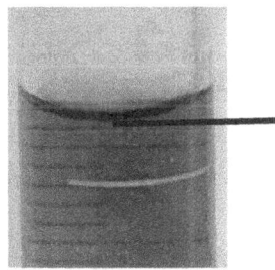

their measurements and calculating standard deviation which is a measurement of precision. The smaller the standard deviation, the more precise is the measurement.

Averages, Standard Deviations, and Significant Figures

All measurements are approximate because there are errors associated with the instrument, the user, and the sample. When you are using natural samples, there is fourth type of error that is due to natural variation of the material because no two cells or organisms are exactly alike. For these reasons, it is important to take every measurement at least three times so you can calculate the average value. In addition, you need to report the standard deviation of your measurements to communicate the confidence of your measurement. Together, the average and standard deviation tell you how many significant figures there are in your value.

When a density is given as 2.170 g/mL, it is communicating that the final digit is significant but somewhat uncertain. By convention, you can assume that any number that lacks a standard deviation is ± 1 for the last digit. In this example, you can assume the value should have been reported as 2.170 ± 0.001 g/mL and that the actual value might be

2.171 or 2.170 or 2.169. Even though the last digit is slightly uncertain, we say this number has four significant figures. Similarly, the number 13.546 has five significant figures. Keep in mind that you need to have a very good instrument, very good user training, and very pure samples to obtain more than three significant figures for any measurement.

When you collect real data and want to report what you found, you need to report the answer using the correct number of significant figures because you want to communicate to the reader that you know which digit has the uncertainty. For instance, let us say you measured the mass of 11 supposedly identical samples and obtained the following results: 20, 24, 26, 24, 30, 30, 24, 26, 20, 24, and 24 g. When you punch these numbers into a calculator, it will give you the average as 24.72727 g and the standard deviation as 3.25856 g. However, it is incorrect to report the result as 24.72727 ± 3.25856 g because both values have too many significant figures. The correct way to report your average and standard deviation is to first round off the standard deviation to one digit because it is telling you about the uncertainty. In this case, the standard deviation should be rounded to 3 grams. Next, you round off the average so that the standard deviation indicates the digit with the uncertainty. In this case, the average should be reported as 25 ± 3 g. Another way to think about it is that your measurements only had two significant figures so the average should only have than two digits.

III. **Community Connection**

Why is measurement important? How many people do you know that add a little of this or that when they are cooking? Others use measuring cups and spoons to be more precise in their cooking.

Measurement is also important for medication, as in a diabetic person who gives themselves their own injections of insulin. Review the community topics (Appendix B), to identify other possible community connections from your own history with this lab.

As important as accurate and precise measurements are, the ability to read measurements while wearing your safety equipment is important as well. This lab allows you to practice using different glassware, identify the equipment that provides the more accurate and precise volume measurements, and practice computational skills you will use in subsequent investigations.

IV. **Safety**

1. Always wear safety goggles; practicing to read while using eye protection in this lab will help you in subsequent labs.

V.   Materials

Equipment	**Samples and Reagents**
Beakers, 50 mL, 100 mL, & 250 mL	Tap water
Erlenmeyer flasks, 125 mL and 250 mL	
Graduated cylinders, 10 mL, 50 mL, 100 mL	
Pipettes and eyedroppers	
Electronic Balance	
Scientific Calculator or Internet Access	

VI.   Procedures

Accuracy

1. Select two beakers, 50 mL and a 250 mL, two Erlenmeyer flasks, 125 mL and a 250 mL, and a 100 mL Graduated cylinder.
2. Fill the 250 mL beaker to the top measurement marking. Record the amount of water in your beaker, using the marking on the beaker in table 1.
3. Pour 50 mL into the 50 mL beaker, using the markings on the beaker to guide you. Record the amount of water in table 1.
4. Pour that 50 mL into a 100 mL graduated cylinder. Record the volume below.

Data Table 1: Beakers

Equipment	Amount of water, mL
Beaker, 250 mL	
Beaker, 50 mL	
Graduated cylinder, 100 mL	

5. Fill the 250 Erlenmeyer flask to the top measurement marking. Record the amount of water, using the flask's markings in table 2.
6. Pour 50 mL into a smaller flask, using the markings on the flask to guide you, Record the amount in table 2.
7. Pour that 50 mL into a 100 mL graduated cylinder. Record the volume below.

Date Table 2: Erlenmeyer flasks

Equipment	Amount of water, mL
Erlenmeyer flask, 250 mL	
Erlenmeyer flask, 50 mL	
Graduated cylinder, 100 mL	

8. Fill a transfer pipette with water and empty it in to a 10 mL graduated cylinder. Record this water volume table 3.

Trial	Volume, mL
1	
2	
3	
4	
5	
Average Volume	
Standard deviation	

9. Average the five measurements and calculate the standard deviation using a calculator, Excel, EasyCalculation.com (https://www.easycalculation.com/statistics/standard-deviation.php), or Calculator.net, (http://www.calculator.net/standard-deviation-calculator.html).

10. Using the electronic balances, measure the weight (mass) of _**two different empty, dry 250 mL**_ beakers three times.

	Mass, g	
Trial	Beaker$_1$	Beaker$_2$
1		
2		
3		
Average mass		
Standard Deviation		

11. Calculate the average and standard deviation as above in #9.

12. Using the electronic balances, measure the mass of a 10 mL graduated cylinder, three times, drying between each trial. Remember to measure at the bottom of the meniscus.

Graduated Cylinder Mass (g)			
Empty	**Filled 10 mL water**	**Mass of water only (g)**	
			Average Mass of water (g)
			Standard Deviation of water mass

13. Calculate the average mass of 10 mL of water. First subtract the weight (mass) of the empty graduated cylinder from the cylinder + water: You may use the average for this.

14. Density is the measurement of the amount of a substance (mass) In a given volume. Calculate the mass of the 10 mL of water:

$$\frac{average\ mass\ of\ 10\ mL\ water}{10\ mL} = \text{Density of water.}$$

15. Clean and return your glassware, return or discard left over materials according to your Instructor, and leave your lab area clean.

VII. Conclusions

Answer on the student report form for experiment 1 (Appendix D).

2. DENSITY

I. Objectives

1. To measure mass of irregularly shaped materials.
2. To measure the volume of irregularly shaped dry matter.
3. To calculate the densities of native seeds, both bulk and true seed density.

II. Facts to Know

Density

The density of an object is equal to its mass divided by its volume:

$$density = mass/volume,$$

which can also be written using symbols as:

$$\rho = m/V, \text{ where } \rho \text{ is the Greek letter rho.}$$

Different sets of units are used to specify density. Here is how they are related:

$$1{,}000 \text{ kg/m}^3 = 1{,}000 \text{ g/l} = 1 \text{ g/cm}^3 = 1 \text{ g/mL.}$$

Density is not only useful for measuring purity and predicting buoyancy but also for useful things such as packaging requirements. It is used to determine purity because pure materials have uniform and characteristic density (Table 1). Buoyancy is the ability to float in liquid or air. A material with lower density will float above one with higher density. With regard to the design of packaging materials, the mass of the material determines how strong the packaging material needs to be, whereas its volume determines the size of the package.

Many commercial products have Irregular shapes so that we need to distinguish **bulk density** (ρ_{bulk}) from **mass density** (ρ_{mass}). If we consider objects such as dried peas, their rounded shapes cause them to pack together in a way that traps a significant volume of air in the container. This increases the volume of the material even though it does not affect the mass very much. Therefore, **bulk density** is the mass of the material that fills a container divided by the volume of that container (reasonably easy to measure), while **mass density** is the mass of the material divided by its volume but excluding the trapped air (a bit more complicated to measure).

When dealing with natural materials like peas and beans, we need to make a further distinction about **mass density** because natural materials have surface pores that are easily penetrated by water. **True seed density** ($\rho_{true\ seed}$) is the type of mass density in which the mass of the seed is divided by its volume, *excluding* open pores. This volume is measured by the displaced water volume method. You place water in a cylinder, record its volume, add the natural material, and record the volume again. The difference between the recorded volumes is the displaced volume. In this case, the water enters the seed's

Table 2.1. Mass densities of pure materials[a]

Material	Density (ρ_{mass}, g/cm^3)
Helium	0.000179
Oxygen	0.0014
Dry Air	0.0012
Styrofoam	0.075
Ethanol	0.789
Methanol	0.791
Water	1.000
Glycerol	1.261
Teflon	2.170
Diamond	3.500
Niobium	8.570
Mercury	13.546
Gold	19.320

[a]Data is from *CRC Handbook of Chemistry and Physics, 64th Edition*, 1983–1984 and *The Engineering Toolbox* at http://www.engineeringtoolbox.com/density-solids-d_1265.html.

surface pores to give a true measure of the seed volume. **Envelope seed density** ($\rho_{envelope\ seed}$) is the type of mass density in which the mass of the seed is divided by its volume, *including* the pores. This volume is measured by placing the seeds in mercury, a non-wetting liquid that does not enter the pores, and then determining the volume of displaced liquid. We will not measure envelope seed density because mercury is toxic. Nevertheless, envelope seed density has become an important parameter to know with the advent of harvesting machinery that collects seeds from their pods or stalks using jets of gas and because of the increasing importance of seed ventilation and drying for storage. To avoid toxic mercury, sensitive and expensive machines have been developed that measure envelope seed density with very high precision.

Although seeds and grains have lower water content than many natural materials (Table 2.2), the moisture content varies a great deal depending on how they are stored. In fact, seed moisture content can be as low as 2% or as high as 30%. Therefore, if you want to compare your measurements to those of other researchers, you need to measure this parameter. The most accurate way to measure moisture content is to determine their mass before and after drying. Unfortunately, we can't wait for a few days while the seeds dry so we won't be measuring moisture content.

Table 2.2. Bulk and true seed densities of cereal grains and legumes[a]

Grain/Legume Type	Moisture Content (% water)	Bulk density (ρ_{bulk}, kg/m^3)	True seed density (ρ_{true}, kg/m^3)
Black-eyed pea	7.3	801 ± 2	1444 ± 1
Chickpea	3.0	868 ± 2	1400 ± 2
Corn	5.6	804 ± 1	1280.7 ± 0.2
Pinto Bean	7.2	841 ± 3	1435 ± 6
Popcorn	4.3	974 ± 3	1398.6 ± 0.3
White Rice	4.8	838 ± 1	1466.2 ± 0.2

[a]Data is from Fasina (2010) *Trans. Am. Soc. Agr. Biol. Eng. 53*, 1223–1227. The values are means ± standard deviations. Compare these to the bulk density averages given by *The Engineering Toolbox* at http://www.engineeringtoolbox.com/foods-materials-bulk-density-d_1819.html.

III. Community Connection

Density is a unique measurement of a known substance. For example, one can identify pure gold based on its density. Food is more complicated but knowing the density of foods may come in handy. Review the community topics (Appendix B) to identify possible community connections.

IV. Safety Concerns

1. Wear eye protection at all times.

V. Materials

Equipment	Samples and Reagents
Electronic balances	Variety of dried native corn kernels
Graduated cylinders, 3-6 100 mL	Popcorn kernels (seeds)
Weighing paper or boats	Dried beans: pinto, fava, lima, green
Internet or scientific calculator	Dried squash seeds: acorn, butternut

VI. Procedures:

We are thankful for the bounty of food that we are able to use the corn, beans, and squash seeds in this lab.

A. Bulk Density

1. Collect your equipment and materials, including your safety gear. Each person or group will select one corn sample, one dried bean sample, and one squash seed sample.
2. Turn on the electronic balance and tare to 0.0 g. Mass an empty, dry weigh boat. Record this value in the data table.

3. Add the dry sample to a dry 100 mL cylinder until it is about one-fourth full. Shake lightly to remove excess air and allow seeds to pack tighter. Record the volume from the very top of your sample.
4. Pour dry contents into your massed boat. Weigh the "sample + weigh boat" and record that mass. The difference between the "sample + weigh boat" and "empty boat" is the mass of the sample.

Bulk Density Data
Name of sample: _____

		Trial # 1	2	3	Average	Standard Deviation
Mass (grams)	**Sample + Weigh Boat**				—	—
	Weigh Boat				—	—
	Difference					
	Bulk Volume (mL)					
	Bulk Density (g/mL)	-	-	-		

Bulk Density Data
Name of sample: _____

		Trial # 1	2	3	Average	Standard Deviation
Mass (grams)	**Sample + Weigh Boat**				—	—
	Weigh Boat				—	—
	Difference					
	Bulk Volume (mL)					
	Bulk Density (g/mL)	-	-	-		

Bulk Density Data

Name of sample: _____

		Trial #			Average	Standard Deviation
		1	2	3		
Mass (grams)	Sample + Weigh Boat				–	–
	Weigh Boat				–	–
	Difference					
	Bulk Volume (mL)					
	Bulk Density (g/mL)	-	-	-		

5. Calculate the average standard deviation *(see below)* of the mass and volume of your sample.
6. Calculate bulk density by dividing the mass with the volume.
16. Calculate the average bulk density and standard deviations of the three trials, using the same tools we used in Lab 1 (using a calculator, Excel, EasyCalculation.com (https://www.easycalculation.com/statistics/standard-deviation.php), or Calculator.net, (http://www.calculator.net/standard-deviation-calculator.html).
7. 8. Repeat steps 2-7 for a total of three trials with each of your three samples.

B. True Seed Density *(also called mass density)*
8. Each person or group will select the same type of corn sample, dried bean sample, and squash seed samples as used in part A to mass and calculate the true seed density.
9. Using the same balance Tare to 0.0 g as needed. Mass an empty boat and record in data table.
10. Place a dry sample on the dry boat that is about the same amount as you used when measuring bulk density.

11. Place your sample + weigh boat on the balance and record its mass. The difference between these values is the mass of the sample.
12. Add tap water to your 100-mL cylinder until it is about half full. Record the volume reached by the meniscus. This is your initial volume.
13. Gently empty the contents from one of your measured weigh boats into the cylinder. Tap the sides gently to remove any air bubbles. Record the volume reached by the meniscus. This is your final volume. The difference between the two volumes is the true volume of your material.
14. Calculate mass (or true seed) density by dividing the mass with the volume and record.
15. Calculate the average mass density and standard deviations of the three trials.
16. Repeat steps 9 – 15 for a total of three trials with each of your three samples
17. Clean and return your glassware, return or discard left over materials according to your instructor, and leave your lab area clean.

Mass (True seed) Density Data
Name of Sample: _____

		Trial # 1	Trial # 2	Trial # 3	Average	Standard Deviation
Mass, g	Sample + Weigh Boat				–	–
	Weigh Boat				–	–
	Difference					
Volume, mL	Final				–	–
	Initial				–	–
	Difference					
	Density (g/mL)	--	-	-		

Mass (True seed) Density Data
Name of Sample: _____

		Trial # 1	Trial # 2	Trial # 3	Average	Standard Deviation
Mass, g	Sample + Weigh Boat				–	–
	Weigh Boat				–	–
	Difference					
Volume, mL	Final				–	–
	Initial				–	–
	Difference					
	Density (g/mL)	-	-	-		

Mass (True seed) Density Data
Name of Sample: _____

		Trial # 1	Trial # 2	Trial # 3	Average	Standard Deviation
Mass, g	Sample + Weigh Boat				–	–
	Weigh Boat				–	–
	Difference					
Volume, mL	Final				–	–
	Initial				–	
	Difference					
	Density (g/mL)	-	-	-		

VII. **Conclusions**:

Answer on the student report form for experiment 2 (Appendix D).

3. CHOCOLATE DENSITY

I. Objectives:
1. Determine the volume of a chocolate bar
2. Determine whether changing the volume a chocolate bar affects its density.

II. Facts to Know

Density is the term used to describe the relationship between the mass of an object and its volume. Under constant conditions of temperature and pressure, the density of any substance is always constant. Gold has a density of 19.3 g/cm³, the average rock has a density of 3 g/cm³, and water has a density of 1 g/mL. 1 g/mL means that 1 mL (or cm³) of water has a mass of 1 gram. In this lab, you will be finding the density of a chocolate bar. The density of substances can be determined by measuring its mass and volume and using the following equation:

$$\text{Density} = \frac{\text{Mass}}{\text{Volume}} \quad \text{or} \quad D = \frac{M}{V}$$

III. Community Connection

Chocolate bars today are not how chocolate consumption began. Native to South America, the cacao plant (*Theobroma cacao*) has been used by Indigenous people for about 4000 years. Originally, the pulp from the cacao fruit was made into a fermented beverage, and only later were the seeds used in beverage making. When the consumption of the seeds themselves became common is uncertain, but the first advertisements in the United States occurred in the mid-1800s (Eschner, 2017). However, chocolate consumption soared after the Swiss added sugar and refined the chocolate into a sweeter, solid treat. The sugar content of many chocolate candies and beverages is a concern among many people. Despite the flavonoids (antioxidants) in chocolate, particularly dark chocolate, and the trace minerals of iron, copper, magnesium zinc, and phosphorus, the sugar content offsets any gains in health and nutrition from the flavonoids (The Nutrition Source, 2018). Chocolate also contains caffeine, with the higher content of cocoa solids having a higher content of caffeine.

IV. Safety
1. Wear safety goggles.
2. Do not eat the food used in this lab.

V. Materials

Equipment
Electronic Balance
Metric ruler
Weighing paper
Gloves

Samples and Reagents
Chocolate bars

VI. Procedure

We are thankful for the bounty of food that we are able to use in this lab.

1. **DO NOT** eat the chocolate bar used in this lab.
2. Unwrap your chocolate bar without breaking it. ***It is very important that it remain in one piece for the beginning of the activity.***
3. Mass the chocolate bar whole. Be sure to zero the balance. Place the chocolate bar on a weigh paper or weigh boat. Record the mass (to the nearest tenth of a gram) of the whole chocolate bar in data table below.
4. Find the volume of the whole chocolate bar using a metric ruler and the equation for the volume of a rectangular solid: v = l x w x h, where v = volume, l = length, w = width, and h = height. Record this measurement (to the nearest tenth) below.
5. Calculate and record the density (to the nearest tenth) of the whole chocolate bar using the density equation.

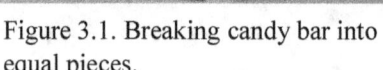

6. Carefully break the whole chocolate bar along the center indentation into two equal pieces (Figure 1). Set one half aside.
7. Repeat procedures 3-5 to find the density of ½ of the bar. Record your answer in the data table to the nearest tenth.

Figure 3.1. Breaking candy bar into equal pieces.

8. Carefully break this half along its center indentation (Figure 2) so that you now have 2 quarters of the original bar. Set one of the quarters aside.
9. Calculate the density for one quarter using step 3 - 6. Record your answer (to the nearest tenth) below.
10. Carefully break one quarter into thirds along the indentations (Figure 3). Set aside the remaining chocolate.

Figure 3.2. Breaking ½ candy bar into equal pieces.

11. Calculate the density of the small piece of the chocolate bar using steps 3 - 6. Record your answer below to the nearest tenth g/cm³.

12. Clean and return your glassware, return or discard left over materials according to your instructor, and leave your lab area clean.

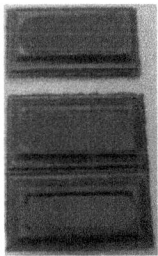

Figure 3 3. Breaking off one section.

Data Table:

	Mass (to the nearest 0.1 g)	**Length** (to the nearest 0.1 g)	**Width** (to the nearest 0.1 g)	**Height** (to the nearest 0.1 g)	**Volume** (to the nearest 0.1 cm³)	**Density** (to the nearest 0.1 g/cm³)
Whole Bar						
Half Bar						
Quarter Bar						
Twelfth Bar						

VII. Conclusions:

Answer the questions on the student report form for experiment 3 (Appendix D).

VIII. References

1. Eschner, K. (2017). A Brief History of Chocolate in the United States. Smithsonian.com Retrieved from https://www.smithsonianmag.com/smart-news/brief-history-chocolate-united-states-180964827/.

2. The Nutrition Source (2018). Dark Chocolate. Harvard T.H. Chan School of Public Health. Retrieved from https://www.hsph.harvard.edu/nutritionsource/dark-chocolate/.

4. LIQUID DENSITY

I.   Objectives

    1.   Determine the densities of several liquids.

    2.   Determine if temperature affects the density of a liquid.

II.   Facts to know

    Why do objects that are the same size sometimes have different weights? The answer has to do with their *density.* An object's density is determined by comparing its mass to its volume. If you compare a rock and a cork that are the same size (they have equal volume), which is heavier? The rock is denser, then, because it has more mass in the same volume - this is due to the elements, molecules, and compounds that make it up.

    The following four experiments investigates characteristics of density in liquids. The first experiment, called "sink or swim," has a surface tension component, which can interfere with your density observations. Make sure the objects are floating due to density, not surface tension.

    The fact that different liquids have varying densities is useful for cleaning up liquid spills, mixing different liquids, and for separating liquids.

III.   Community Connection

    Areas of interest within your community may encompass the water quality of your nearest aquifer, the Missouri River, or another major river, local, smaller river or stream, well water, ponds or lakes. Soil quality and water run-off is another topic of interest that may relate to liquid density. Oil pollution from spills from pipelines or tankers, oil contamination from drilling and extraction, or even from daily household uses is another area of concern and research.

    In the chemistry community connection, "Life Flows", several Native American tribes water stories are shared. The chemistry of a water as well as some properties of water is presented to increase understanding of water's unique characteristics and importance. Then, a variety of problems where this knowledge may be applied is given. Perhaps you would like to explore further? For example, does density play a role in water pollution? How can we use density to clean up water pollution? How does the density of water affect agriculture and crop growth?

IV.   Safety

    1.   Always wear safety goggles.

    2.   Gloves and a lab coat may prevent stains from food coloring.

V.   **Materials**

Equipment

Beakers, 150-250 mL,

Beaker, 400-600 mL

Several small objects: raisins, paperclips,
pennies, small corks, etc.

Spoons and forks

Graduated cylinder, 10 mL and 100 mL

Pipettes

Samples and Reagents

Water, distilled and tap

Corn syrup

Vegetable Oil

Food Coloring, blue, green, & red

Table sugar

Table salt, NaCl

VI.   **Procedures**

A. **Sink or Swim: Surface tension or Density?**

1. Record the small objects you are using in the data table below.
2. Record your prediction of whether the object will sink or swim (float).
3. Label the beakers, 1, 2, and 3. Pour 50 ml of water into beaker 1, 50 ml of corn syrup into beaker 2, and 50 ml of vegetable oil into beaker 3. Record the brand and type of corn syrup and vegetable oil used the heading of the data table.
4. Gently set one of your small objects at a time into each beaker. Does it sink or float? Record what happens to each small object in each beaker.
5. Remove the small objects from the beakers with a spoon or fork. Do keep your three beakers with their liquids as we will use them again.

Sink or Swim (Surface Tension) Data Table

Small Object Tested	Beaker 1 Water		Beaker 2 Corn Syrup		Beaker 3 Vegetable Oil	
	Prediction	Actual	Prediction	Actual	Prediction	Actual

B. Mix it up (Relative Liquid Density)

1. Based on your results from experiment A, predict which liquid is the densest and which is the least dense:
2. Place a few drops of food coloring into the beaker of water so you will be able to tell it apart from the other liquids.
3. Carefully pour each of the liquids into a 400 or 600 mL beaker. While these liquids are settling, do Experiment C.
4. After the liquids have settled, which liquid is at the bottom (densest), top (least dense), and in the middle.
5. Clean and return your glassware, return or discard left over materials according to your instructor, and leave your lab area clean.

Mix it up (Relative Liquid Density) Data Table

Liquid	Predict which liquid is the most dense, least dense, and middle density	Actual density most dense, least dense, and middle density
Corn Syrup		
Vegetable Oil		
Water		

C. Hot and Cold (Temperature effects on water density)

1. Fill two beakers with 50 ml of water. Put several drops of blue food coloring in one beaker, and several drops of red In the second.
2. Add a handful of ice to the blue water and put it in the refrigerator's freezer for 10 minutes. Measure and record the temperature.
3. Put the red water solution r in the microwave for 2 minutes. Alternatively, you may heat water using a hotplate; it does not have to boil. Measure and record the temperature.
4. Remove the blue water solution from the fridge and the red beaker from the microwave (or off the hotplate).
5. Pour approximately 5mL of the blue water into the 10-ml graduated cylinder or a narrow glass container.
6. Slowly using a pipette, add the red water, a few drops at a time, and watch what happens. (This part may take a little practice--if you add the red water too fast you will force the colors to mix. Hold the pipette near the surface of the water or let the drops roll down a side of the container and keep trying until you get it!)

7. Record your observations.
8. Clean and return your glassware, return or discard left over materials according to your instructor, and leave your lab area clean.

Hot and Cold (Temperature effects) Data Table

Water Sample	Temperature, °C	Observations
Blue water		
Red water		
Mixture		

D. Salty or Sweet (Salt and Sugar effects on water density)
1. Determine whether salt or sugar will increase or decrease water density. *(Circle your prediction)*
   a. Adding salt to water will make water denser.
   b. Adding salt to water will make water less dense.
   c. Adding sugar to water will make it more or less dense
   d. Sugar water will be denser salt water.
2. Fill three beakers with 50 ml of water. Add food coloring to make blue, red, and green water.
3. Add 2 teaspoons of salt to the red beaker and stir until the salt is dissolved.
4. Add 2 teaspoons of sugar to the blue water and stir until it is dissolved.
5. Pour some of the red (salty) water into the 100 mL graduated cylinder. Using the pipette, slowly add the blue (sugar) water one or two drops at a time. Record which sinks to the bottom and which floats on top.
6. Add the green (pure) water drop-by-drop to the other two and record what happens.
7. Dispose of your waters, rinse, and add 50 mL of water to each of the beakers, adding food coloring to make blue, red, and green water.
8. Repeat steps 3-6, but this time add 10 grams of salt to the red beaker and 10 grams of sugar to the blue water. Stir both until the salt and sugar are dissolved. Use a different 100 mL graduated cylinder so you can compare the two methods.
9. What do your graduated cylinders with the waters look like? Take a photograph of your graduated cylinder after the waters have settled, about 5-10 minutes, or draw a picture of the two graduated cylinders labeling clearly.
10. Clean and return your glassware, return or discard left over materials according to your instructor, and leave your lab area clean.

11. Go back to experiment 2, and record the conclusions from observing the layers of liquids in the beaker.

V. Conclusions

Answer the questions on the student report form for experiment 4 (Appendix D).

5. PERIODIC TABLE VIDEO

I. Objectives

1. To learn about the chemical and physical characteristics of 14 elements.
2. To familiarize yourself with the Periodic Table

II. Background Information

Video journalist, Brady Haran, at the University of Nottingham, launched the "Periodic Table of Videos" series in 2008. Chemistry Professor, Martyn Poliakoff, is the narrator. After making short videos about all 118 elements (2 to 10 min each), the team began updating videos for some of the elements and creating entirely new series about molecules, numbers, physics, etc. The participants in this video documentary are professional educators and journalists. Their goal is to bring interesting chemistry knowledge to the general public. These videos are free to the public through the following links:

http://www.periodicvideos.com/

http://www.youtube.com/user/periodicvideos

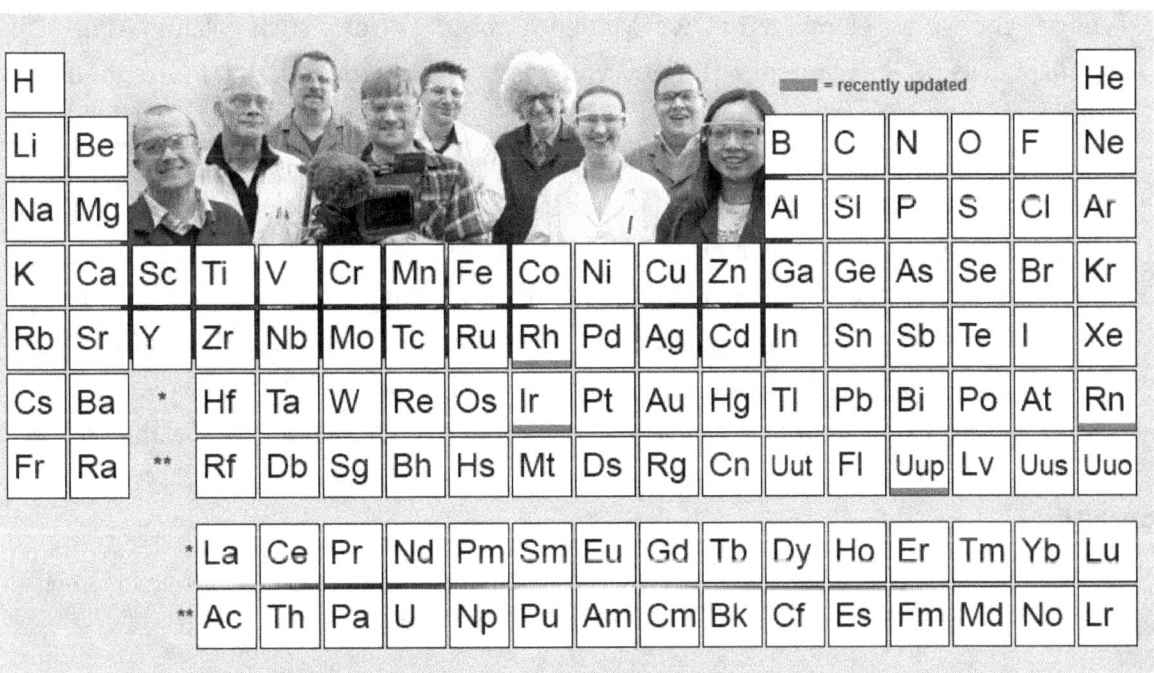

The "Periodic Table of Videos" website (www.periodicvideos.com)

Image reproduced with permission from Brady Haran (the one in the plaid shirt)

III. Community Connection

Knowledge of the individual elements and their properties provides a foundation for our understanding of chemicals, their properties, and their reactions. This may assist in understanding concerns of the quality and contamination of our atmosphere, foods, soils, and water.

Review the community topics (Appendix B) to identify possible community connections.

IV. Safety Considerations

1. Do not sit too close or too far from the television screen; you might strain your eyes
2. Do not try to repeat the demonstrations shown in the videos. Some of them involve high- energy explosions or very hazardous materials.

V. Materials and Equipment

Television monitor and/or computer with Internet connectivity

VI. Procedures:

1. Watch the videos for elements 1 through 11 (hydrogen through sodium) during lab.
2. Choose three additional element videos that interest you to watch on your own. You might choose an element that we will learn about in class, such calcium (Ca), phosphorus (P), or potassium (K). Or, you might watch one of the more popular videos. Four of the most popular videos are for H, He, O, and F, which you will have already watched. Among the next most popular are K, Cs, Fr, Hg, U, Pu, Fe, Xe, and Pb. Some of these are popular for the demonstrations, while others are popular because they are well-known and/or toxic elements.
3. It is advisable to take notes while you watch the videos or else you will have to watch them again later. You will have to perform some investigations after class to find the information regarding properties or everyday use of these elements. Most of this extra-laboratory material can be found in your chemistry textbook.

VII. Conclusions:

Answer questions on student report form (Appendix D).

6. WATER QUALITY ANALYSIS

I. Objectives

1. Develop an understanding of what is meant by "water quality" and how quality is tested.
2. Determine which inorganic ions are present in the water samples based on qualitative tests, pH, and conductivity and how these relate to water quality.
3. Practice testing protocols for a variety of water samples
4. Relate water quality with the appropriate community topics listed in Appendix C.

II. Facts to Know

How safe is your drinking water? Let's find out by testing a variety of water samples from your wells, rivers, and communities.

Drinking water from your tap is never free of chemicals in the ionic form. Most municipalities treat their water with chlorine to protect against common bacteria. Fluoride is added for strengthening bones and teeth. Chemicals from the rock type where the water originates and even ions from your plumbing can dissolve into your water. If you obtain your drinking water from a well, then anything that is in the surrounding ground will be in the groundwater feeding your well. Irrigation water may flow into your drinking water source and increase the concentration of nitrates and phosphorus from fertilizers, as well as pesticides added to the crops. Most drinking water is slightly acidic due to dissolved CO_2, which forms carbonic acid, a weak acid.

Qualitative analysis is the identification of a sample's components. This contrasts with quantitative analysis, which seeks to determine the amounts of each component. The best qualitative analysis method should be able to identify a single type of ion even in the presence of other similar ions. For instance, the ideal analysis would be able to establish the presence of a small amount of sodium ion in the presence of a high concentration of potassium ions. Unfortunately, this is not always possible because sodium and potassium have similar chemical properties. It is only possible to distinguish sodium and potassium by conducting a series of analytical experiments. After that, it is necessary to confirm the identity of each ion by isolating them from the solution.

In this experiment, you test for a variety of inorganic ions from your water samples. The two most common reactions used to confirm the presence of ions in solutions are called precipitation and complexion. Precipitation reactions occur when certain ions are added to a test solution and a solid is formed. Since it is already well known which cations form insoluble complexes with which anions, it is a simple matter to select test solutions that can distinguish between many salt solutions. Many textbooks list the common soluble and non-soluble ions as well as the solubility rules. Complexation reactions are when cations form covalent bonds with one or more ligands. Most of these have distinct colors and are very soluble. Many transition metal ions form color complexes.

Precipitation reactions depend on soluble ions in a solution that react to form a precipitate, or solid. These reactions are visually simple to identify because a solid forms in the reaction container. For example, an aqueous solution (aq) of silver nitrate added to an aqueous solution (aq) of potassium chloride will form a white precipitate (s) of silver chloride and the potassium and nitrate will stay in solution as ions:

$$AgNO_{3(aq)} + KCl_{(aq)} \rightarrow [Ag^+_{(aq)} + NO_3^-_{(aq)} + K^+_{(aq)} + Cl^-_{(aq)}] \rightarrow AgCl_{(s)} + K^+_{(aq)} + NO_3^-_{(aq)}$$

Reagents   Ions in solutions   White Solid Product

Complexation reactions are not as straightforward as precipitation reactions. These reactions form a "complex", hence their name. The complex forms using a special type of covalent bond, called a co-ordinate or dative covalent bonds. The complex ion has a metal ion in the center and is surrounded by a number of other molecules or ions. A complex ion involves more than one species, $[FeSCN]^{2+}$, and still carries an overall charge, this it is a complex ion while in solution. Complex ions frequently form around a transition metal cation, is dependent upon concentration, and is usually indicated by a color change in the solution. Most complex ions are brightly colored as well. For example, the species given previously, $[FeSCN]^{2+}$, forms from adding a potassium thiocyanate, KSCN, to a solution containing iron (III) ion. This complex ion will be a deep blood red.

$$Fe^{3+}_{(aq)} + KSCN_{(aq)} \rightarrow \{[Fe(H_2O)_6]^{3+} + K^+_{(aq)} + SCN-\} \rightarrow [FeSCN]^{2+}_{(aq)} + K^+(aq)$$

Reagents   Ions in solution   Blood red color

                 Brown color   No precipitate

(Due to iron-water complex ion)

III. **Community Connection**

The community connection, Life Flows, provides background information about water and water chemistry as well as posing questions that could be pursued for a final project. It also was written with the community topics (Appendix B) in mind, specifically looking at community and individual's health. Review the community topics (Appendix B) to identify possible community connections.

It is important to know what ions are present in the water you drink. Some ions are harmful at high concentrations. For instance, 80 communities in Nebraska have wells containing arsenic and/or uranium at levels that exceed the Federal Government Standards. University of Nebraska-Lincoln released results of studies of two major aquifers in 2015. The aquifer underlying the Great Plains shows contamination with naturally occurring uranium (University of Nebraska-Lincoln, 2015). Some water contaminants, such as nitrates from excess fertilizer run-off, may affect the pH of the water which in turn may affect the solubility of other ions in soils.

IV. Safety Considerations

1. Wear safety goggles.
2. Avoid skin contact with the chemical solutions by wearing gloves.
3. Replace caps and covers of vials and bottles immediately after use to prevent contamination.
4. Wash your hands (and arms if you splash) thoroughly after you have been working with water of unknown quality.

V. Materials

Equipment	Samples and Reagents*
Bottles, collecting water samples	Water samples
Graduated cylinders	pH paper or universal indicator
Well plates	Iron Test: 0.5 M potassium thiocyanate, KSCN
Thermometer	Chloride test: 0.1 M silver nitrate, $AgNO_3$
Pipettes, transfer	Calcium Test: 0.1 M sodium carbonate,
Test tubes & test tube rack	$Na_2C_2O_4$ & 1.0 M acetic acid, $HC_2H_3O_2$
Conductivity probe or meter	Ammonium Test: 1.0 M sodium hydroxide,
pH meter	NaOH & red litmus paper
Beaker, 250 mL	Nitrate Test: 1.0 M sodium hydroxide and
Spatula, small	$1cm^2$ piece of aluminum foil
Hot plate	Sulfate test: 0.1 M $BaCl_2$
	$\geq$70% Isopropyl Alcohol
	Phosphate test: 0.1 M ammonium molybdate,
	$(NH_4)Mo_7O_{24}\cdot4H_2O$, ascorbic acid crystals
	Alkalinity and hardness test strips[a]
	Pesticide test strips[b]
	Assorted petrifilm, Aerobic, E. Coliform

*You may prefer to use water testing kits instead. If so, then follow their instructions for testing. Probeware would be another option for water testing.

[a]These may be purchased anywhere chemicals for home pools and hot tubes are sold.

[b]These are available from chemical laboratory suppliers and test for atrazine and simazine as well commercially where you can purchase gardening equipment and water testing kits.

VI. Procedures

1. Obtain equipment and materials for testing water sample.
2. Measure the temperature of your water sample when you collect it and record in the data table.
3. Measure the **pH** of your sample with pH meter, paper, or universal indicator.
   a. Follow the instructions for the **pH meter**.
   b. If using **pH paper**, fill a well plate $\frac{2}{3}$ full with the water to be tested. Dip a strip of pH paper into the water, determine the pH with the color chart provided, and record.
   c. If using **Universal pH indicator**, fill a well plate $\frac{2}{3}$ full with the water to be tested. Add 1 drop of Universal pH indicator and mix with the plastic spatula's flat end.

   Compare the color that **_immediately_** appears with this list:

pH	Color	pH	Color
1	Cherry red	6	Yellow
2	Rose	7	Yellow-green
3	Red-orange	8	Green
4	Orange-red	9	Blue-green
5	Orange	10	Blue

4. To test for **Ammonium ion,**
   a. Measure 2 mL water sample into a test tube.
   b. Add 3 drops of dilute sodium hydroxide.
   c. Place in a hot water bath but do not let the water sample-NaOH mixture boil.
   d. Test the pH of the gas given off from the mixture by holding a moisten red litmus paper near the top of the test tube; it should turn blue if NH_4^+ or NH_3 is present. There should be the odor of ammonia.
   e. This reaction is written as $NH_4^+{}_{(aq)} + NaOH_{(aq)} \rightarrow NH_{3(g)} + Na^+{}_{(aq)} + H_2O_{(l)}$.

5. To test for **Chloride ion,**
   a. Measure a 2 mL of your water sample into a clean test tube.
   b. Add 3 drops of 0.1 M $AgNO_3$ solution.
   c. Mix thoroughly by swirling.
   d. IF chloride ions are present, a white precipitate will develop due to the formation of AgCl. Allow 5 minutes for full color development. This reaction is given in the 'Facts to Know' Section.

6. To test for *Iron ion*,
    a. Measure a 2 mL of your water sample into a clean test tube.
    b. Add 3 drops of 0.5 M KSCN solution.
    c. Mix thoroughly by swirling.
    d. IF iron ions are present, a deep red color will develop to the formation of the iron (III) thiocyanate complex, $Fe(SCN)^{2+}_{(aq)}$. Allow 2 minutes for full color development. This reaction is given in the 'Facts to Know' Section.

7. To test for *Calcium ion,*
    a. Measure a 2 mL of your water sample into a clean test tube.
    b. Add 3 drops of 1.0 M $CHCH_3O_2$, does not take part in the reaction but helps sodium oxalate disassociate.
    c. Add 3 drops of 0.1 M $Na_2C_2O_4$.
    d. Mix thoroughly by swirling
    e. IF calcium ions are present, a yellowish white solid (precipitate) will form, calcium oxalate (CaC_2O_4), the main constituent of kidney stones
    f. This reaction is written as $Ca^{2+}(aq) + Na_2C_2O_{4(aq)} \rightarrow CaC_2O_{4(s)} + 2Na^+_{(aq)}$.

8. To test for *Nitrate ion,*
    a. Measure 3 mL of water sample in the graduated tube.
    b. Add 2 mL of 1 M NaOH and swirl gently to mix.
    c. Add a 1 cm^2 piece of aluminum foil to the mixture, the aluminum is reduced the nitrate ion to form the ammonium ion which then reacts with the sodium hydroxide.
    d. Warm carefully in the hot water bath-do not let the mixture in the test tube boil.
    e. IF nitrate is present ammonia odor is detected. A moistened red litmus paper strip will turn blue if inserted into the top of the test tube.
    f. This reaction is written as $NH^+_{4(aq)} + NaOH_{(aq)} \rightarrow NH_{3(g)} + Na^+_{(aq)} + H2O_{(l)}$.

9. To test for *Phosphate ion,*
    a. Set up a hot water bath by filling the 400 mL beaker 300 mL of tap water.
    b. Measure 15 mL water sample in into a test tube.
    c. Add 1 mL of 0.1 M $(NH_4)Mo_7O_{24}\cdot4H_2O$.
    d. Add a small spatula measure of ascorbic acid crystals ($C_6H_8O_6$), assists in disassociation of the ammonium molybdate.
    e. Mix thoroughly and place in a hot water bath. Bring the mixture to a boil and then allow to cool.
    f. IF phosphate ions are present, a deep blue color will form due to the formation of the phosphomolybdate anion which reduces even more to form a mixed-valance complex, which gives the deep blue color.

g. This ion reaction is written as

$$PO_4^{3-}{}_{(aq)} + (NH_4)Mo_7O_{24}\cdot4H_2O_{(aq)} + C_6H_8O_{6(s)} \rightarrow [P(Mo_4)^{5+}(Mo_8)^{7+}O_{40}]^{7-}{}_{(aq)}$$

10. To test for **Sulfate ion**
    a. Measure a 2 mL water sample in to a test tube.
    b. Add 3 drops of 0.1 M $BaCl_2$.
    c. Mix thoroughly by swirling.
    d. IF sulfate ions are present, a white precipitate of $BaSO_4$ will form.
    e. This reaction is written as $SO_4^{2-}{}_{(aq)} + BaCl_{2(aq)} \rightarrow BaSO_{4(s)} + 2Cl^-{}_{(aq)}$.

11. To test for pesticide,
    a. Measure 10-15 mL water sample into graduated tube.
    b. Dip test strip into water and read results.

12. If there are other tests you wish to perform, and we have time and the materials, add here, following the protocol provided by your TA/instructor. Add in at the bottom of the data table.

13. Screen your water sample for **microorganisms**: Obtain petrifilm from your instructor; sterilize eyedropper or pipette with $\geq$70% Isopropyl Alcohol, then rinse with water sample twice before applying water to the petrifilm surface. Follow the instructions with the petrifilm for dispersal. Place in a safe, warm location to monitor growth over the next week, checking daily. Record results.

14. Clean and return your glassware, return or discard left over materials according to your instructor, and leave your lab area clean.

Data Table

Water Sample Name		Location	
Temperature of water when collected (in field)			
Ion/Chemical Test	Color Observed		Present (+) or absent (-)
pH			pH # =
Ammonia			
Chlorine			
Iron			
Calcium			
Nitrate			

Phosphate		
Sulfide		
Pesticide		

VII. Conclusions

Answer questions on student report form (Appendix D).

VIII. Reference

1. University of Nebraska-Lincoln. (2015, August 17). Two major US aquifers contaminated by natural uranium: Naturally occurring uranium is being mobilized by farm-related pollution. *ScienceDaily*. Retrieved from www.sciencedaily.com/releases/2015/08/150817132508.htm.

PROPOSAL FOR END OF FIRST SEMESTER PROJECT

The last lab is a student project (#13). Students are to connect chemistry, a community connection (Appendix B), and their own personal interests. Possible ideas for projects might be a short story, a film, a photo journal, a play, a costume, beadwork, a dance, a game, an experiment, a research paper … there are no limits as long as you can connect your project with chemistry!

A short proposal of your selected topic and preferred format (i.e. visual media, research, experiment) is due at midterms. Your instructor will provide you with requirements for this proposal.

7. WATER PURIFICATION

I. Objectives

1. To purify a foul water sample.
2. To compare federal water quality standards with your local water quality.
3. To become aware of community topics related to water quality and uses.

II. Facts to Know

A brief review of the water cycle may be beneficial to understand natural water purification. Water evaporates into the atmosphere, returns to the Earth's surface via precipitation, some of the water runs off the surface into streams and lakes or back into the oceans. Some of the water on the Earth's surface infiltrates the ground and may replenish aquifers. Water from aquifers are returned to the surface via springs and wells, artesian and human-drilled.

Liquid water purification occurs naturally through filtration over time. This process depends on chemical absorption and adsorption by soil particles and organic matter, living organisms' uptake of nutrients, and decomposition processes in the soil and in water bodies.

Soils and the associated vegetation and microorganisms play major roles in natural water purification, particularly in wetland and riparian zones. Microorganisms in the soils and water bodies utilizes or decompose chemical and biological contaminants in water.

Water that infiltrates the ground and is not taken up by living organisms or adsorbed by organic matter, percolates through permeable ground layers. Sandy layers increase the movement of the water while clay layers decrease if not stops water movement. Eventually this purified water becomes part of the aquifer systems.

Underlying the Great Plains is the largest aquifer in the United States, the High Plains aquifer. This aquifer provides drinking and irrigation water for South Dakota through Nebraska into northern Texas. Coupled with the Central Valley aquifer which sits beneath the fertile agricultural land of California and the High Plains aquifer, the irrigated cropland provides approximately one-sixth of the annual agriculture revenue in the U. S. A. (University of Nebraska-Lincoln, 2015).

III. Community Connection

Where does your water come from? If you live in a non-urban location, perhaps a well? How clean is that well water? Has It been tested? Is it free of contaminating chemicals and fauna? How do you know if your water is safe to drink?

In urban areas, the water is treated prior to consumption by the populace. Or is it? During spring run-off, the amount of water entering a city's water system may actually overwhelm the facility so that untreated water enters into the water system for consumption without treatment and can create health concerns, particularly bacteria contamination. Usually when this occurs the municipality will issue an alert for residence to either not use their city-provided water or to boil it before use. However, if you are in a rural area, the flooding can impact your previously pristine wells. Residents relying on private wells should have their monitor their wells and have the water tested during and after flooding.

Review the community topics (Appendix B) to identify possible community connections.

IV. **Safety Considerations**
1. Wear safety goggles.
2. Be careful cutting and building your water purification column from the plastic bottles.
3. Do not consume your foul water sample before OR after purification. We have not sterilized the water so it still may contain dissolved chemicals or microorganisms that are unhealthy.

V. **Materials**

Equipment	Samples and Reagents
Plastic bottle, two 2- liter, per group	Foul water sample*
Scissors or knife	Alum
Beakers, 3-400 mL	Fine sand, washed, $\geq$ 1 Cup /group
Spatula or spoon	Coarse sand washed, $\geq$ 1 Cup /group
Filter paper, coffee filter works fine	Small clean pebbles, $\geq$ 1 Cup /group
Rubber bands	Cotton
Timer	$\geq$ 70% Isopropyl alcohol
Ring stand with round clamp	Assorted petrifilm, aerobic, E.
Transfer pipettes or eyedroppers	coli/coliform counting plates

Your instructor has concocted your four-water sample with common household materials; none are hazardous or toxic, but you still will not taste your purified sample.

VI. Procedures

1. Gather your materials and equipment, put on your safety goggles.

2. Cut one of the 2-liter plastic bottles approximately in half as shown to the right. (Figure 7.1)

3. Pour about 300 mL of your foul water sample into one of the 400 mL beakers. Record your observations the sample, describing what you see and smell. Do not taste! You may take a photo of it to add to your observations.

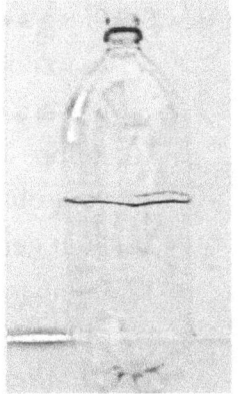

Figure 7.1. Cut your 2-liter body in half.

4. Pour this 300 mL sample into one of the 2-L plastic bottles, place the cap back on snugly, and shake vigorously to **aerate**. Continue aeration process by pouring water back into the 400 mL beaker.

5. Now pour your sample back and forth between the bottle and beaker or two beakers, 10 times. Pour your 300 mL sample in to the bottom half of your cut plastic bottle.

6. Add two tablespoons of alum to the aerated water. Slowly stir the mixture for 5 minutes. Particles will begin clinging together to form **floc** (clumps of sediment and alum), this is called **flocculation**, a type of **coagulation**. The floc becomes heavier as it clumps more and more sediment and will settle to the bottom in a process called **sedimentation**.

7. Allow the water to stand undisturbed in the bottle half for 20 minutes. Observe every 4-6 minutes, recording your observations.

8. Construct a filter from the other half of the cut plastic bottle, using the coffee filter, rubber band, sands, and pebbles. (Figure 7.2).

   a. Attach the coffee filter to the outside neck of the bottle with a rubber band.

   b. Turn the bottle upside down and place in the ring clamp attached to the ring stand.

   c. If you have cotton, place some inside the neck touching the filter paper.

Figure 7.2. Coffee filter, pebbles, and sand filtration.

   d. Pour a layer of pebbles into the bottle. This layer should fill about $\frac{1}{4}$ of the bottle.

   e. Pour the coarse sand next to about half full.

   f. Pour the fine sand on top of the coarse sand layer. The bottle should be about $\frac{3}{4}$ full.

   g. Note: You can add activated charcoal but wash it thoroughly prior to use else your water will have charcoal particles in.

9. **Place a 400 mL beaker under the filter. Wash with distilled water. Pour carefully** so as to not disturb the fine sand layer. You may use a funnel to assist you in pouring. Discard the water you washed your filter with.
10. Before your pour your aerated sample of foul water that you have let settle, record your observations of the sample.
11. Pour gently (decanting) the liquid layer without disturbing the sediment.
12. Compare your foul water sample with your filtered sample. Record your observations.
15. *Optional*: Screen water sample for *microorganisms*: Obtain petrifilm from your instructor; sterilize eyedropper or pipette with $\geq$70% Isopropyl Alcohol, then rinse with water sample twice before applying water to the petrifilm surface. Follow the instructions with the petrifilm for dispersal. Place in a safe, warm location to monitor growth. Record results.
13. Clean up your lab area. Dispose of the sand and pebbles in the container provided. Clean your glassware.

Data Table

Water Sample # _____	Observations
Pretreatment Foul water sample	
Sedimentation sample: Initial	
10 minutes	
30 minutes	
40 minutes	

2 days	
Filtered Sample	

VII. Conclusions

Answer questions on student report form (Appendix D).

VIII. References

1. Center for Engineering Educational Outreach. (2013). *Hands on Activity: Water Filtration Project Make Your Own Water Filters.* Retrieved from https://www.teachengineering.org/activities/view/water_filtration.

2. Home Science Tools. (2018). Water Purification Science Project + Video. Retrieved from https://www.homesciencetools.com/article/water-filtration-science-project/.

3. University of Nebraska-Lincoln. (2015, August 17). Two major US aquifers contaminated by natural uranium: Naturally occurring uranium is being mobilized by farm-related pollution. *ScienceDaily*. Retrieved from www.sciencedaily.com/releases/2015/08/150817132508.htm.

8. SOIL QUALITY ANALYSIS

I.   Objectives

1.   To test soil samples for inorganic ions, pH, pesticides, and conductivity.
2.   To connect community topics related to soil quality and use.

II.   Facts to Know

Plants get nearly all of their nutrients from the soil using their roots. The main exception is the carbon dioxide they extract through vacuoles on their leaves and stems. Since plants need large quantities of nitrogen, phosphorus, potassium, you can enhance the growth rate of your plants by giving them the right amount of each of these nutrients. Plant growth is enhanced when nutrients are present in the right amount in the soil as well as optimum pH levels for nutrient uptake. Once the nutrient content of a soil area is determined, additions of other substances can help adjust the nutrients to that optimum level.

Nitrogen is required in the greatest quantity when plants are growing fastest and is often the "limiting nutrient" during that period. Nitrogen is needed to make many molecules in the plant including proteins called enzymes (the engines of the cells) and chlorophyll (the green- colored molecule that converts the sun's energy into electrical energy to "fix" carbon dioxide). The right concentration of nitrogen in the soil will produce healthy green plants that grow fast. If the soil nitrogen content is too little or way too much, plant growth will be stunted. The nitrogen can be provided in the form of decaying organic matter (all living things have many nitrogen compounds in them), manure (rich in aromatic amines), urine (rich in urea), or fertilizers (such as ammonium nitrate). We will measure the nitrate content in some soil samples.

We will also measure the pH of those soil samples. The availability of essential nutrients and toxicity of other elements can be estimated by the complex relationship they have with the soil's pH. It is most likely that the soil will be acidic, and the pH will be less than 7. Soils that are very acidic have a pH in the range of 4 to 5. At that pH, the calcium ions and magnesium ions tend to be low because these two ions bind weaker to the other molecules in the soil at that pH and they flow away in the run-off. This is bad for the plants because these two ions are important nutrients. Also, at this low pH, the amount of soluble manganese and aluminum ions increase. (Aluminum ions compose as much as 12% of soil.) This is a problem because these ions are toxic to the plants. To counteract the low pH, powdered limestone $CaCO_3$ is added until the pH rises to between 6.0 and 6.5, the optimum pH for solubility of the nutrients and for the growth of nitrogen-fixing bacteria in the soil. Adding too much limestone will raise the pH too much and create its own set of problems.

Since nitrate is not particularly reactive and is certainly not colorful, you will use an "indirect colorimetric assay" to measure the nitrate content of the soil. This means that you will transform the nitrate to something else that can produce a color. We will test for nitrates, phosphorus, potassium, and two pesticides. The chemical reactions with the ions or pesticides will cause a color change or produce a precipitation. Strength of color change may indicate concentration of the ions and pesticides, but we will not have a value for the concentration, only that the ions are present, which is why these are qualitative tests not quantitative tests.

III. Community Connection

Soil quality is important for local gardeners as well as larger agricultural enterprises. Types of soil, nitrogen-phosphorus-potassium (N-P-K) concentrations, and pH affect quality of plant growth and fruit/seed production. The preferred N-P-K ratios are 3-1-2, that is 3%-1%-2% for most vegetable gardens. The N-P-K ratios for commercial agricultural crops varies according to the specific crop. Commercial fertilizers will have a 3-2-1 or a multiple of that ratio on their ingredient label. Most vegetables grow well in soils with a pH of 6.0 – 7.0. Agricultural crops also have optimal soil pH unique to the crop, although many fall within the 5.5-6.5 range.

Review the community topics (Appendix B) to identify possible community connections.

IV. Safety Considerations

1. Wear safety goggles;
2. avoid skin contact with the chemical solutions; flush with water if you do;
3. replace caps and covers of vials and bottles immediately after use to prevent contamination.

V. Materials

Equipment	Materials[a]
Graduated cylinders, 10 mL & 50 mL	Soil samples
Test tubes	Nitrate Test: 1.0 M sodium hydroxide, NaOH & red litmus paper
Beaker, 2- 250 mL	
Pipettes, transfer	Potassium test: 1% w/v sodium tetraphenyl borate, $NaB(C_6H_5)_4$, in 0.01 M NaOH,
Rubber stoppers	
Conductivity probe or meter	Phosphate test: 0.1 M ammonium molybdate,
pH meter, if have access to one	$(NH_4)Mo_7O_{24} \cdot 4H_2O$, ascorbic acid crystals
Trowel or other small digging tool	pH indicator
Clean, dry containers for soil samples	Distilled water
Small spatula	Pesticide test strips*

*These may be purchased at commercial businesses that sell gardening equipment or water testing kits as well as laboratory chemical equipment and materials. They test for atrazine and simazine.

ªProbewear or soil testing kits may be used instead of the chemical solutions.

VI.   **Procedures**

Soil sampling

1.   Clear of the area you want to sample of plants material and other debris.
2.   Using a trowel or large spoon, remove the soil to a depth of about 4 inches (10 cm). Avoid touching the soil with your hands. IF you have a soil pH meter, follow those instructions to test the soil's pH at your sample site. Record for later use.
3.   Place your sample into a clean container (beakers, glass or plastic jars, zip lock bags, muslin sample bags, etc.). Break the clumps with the trowel. Remove any small stones, organic material, wearing gloves. The crumple the sample finely and mix thoroughly.

Soil testing

4.   To test the **pH** in the lab,
     a.   Fill the test chamber or tube with 1 mL (0.5 teaspoon; or to the fill line) of soil from your sample.
     b.   Follow instructions if you are using a **pH meter**.
     c.   Using **test kit chemicals**, dispense the powdered chemical from the pH capsule into the soil sample, add 5 m L water (or to the fill line) with a transfer pipette. Cover tightly with cap or stopper and shake vigorously. (*Do not use your thumb; avoid contamination of sample and touching the chemicals.*)
     d.   Compare the color solution to the pH chart provided and record the pH value of your sample in the data table.
5.   For the remaining tests, first fill 2-L container with 236 mL (1 cup) of soil and 1.2 L (5 cups) of distilled water. (You may modify the amounts as long as you maintain the ratio of 1-part soil to 5 parts of water.)
6.   If your container has a lid or cap, cover tightly and shake to mix, otherwise stir to mix well for at least one minute. Allow the mixture to settle undisturbed for 30 minutes to days. The finer grained your soil, the longer it will need for particles to settle out of the solution.
7.   To test the **nitrogen ion**,
     a.   Measure 3 mL of water sample in the graduated tube.
     b.   Add 2 mL of 1 M NaOH and swirl gently to mix.

c. Add a 1 cm^2 piece of aluminum foil to the mixture, the aluminum is reduced the nitrate ion to form the ammonium ion which then reacts with the sodium hydroxide.

d. Warm carefully in the hot water bath-do not let the mixture in the test tube boil.

e. IF nitrate is present ammonia odor is detected. A moistened red litmus paper strip will turn blue if inserted into the top of the test tube.

f. This reaction is written as $NH^+_{4(aq)} + NaOH_{(aq)} \rightarrow NH_{3(g)} + Na^+_{(aq)} + H2O_{(l)}$.

8. To test for *phosphate ion*,

   a. Set up a hot water bath by filling the 400 mL beaker 300 mL of tap water.

   b. Measure 15 mL water sample in into a test tube.

   c. Add 1 mL of 0.02 M $(NH_4)Mo_7O_{24}\cdot4H_2O$.

   d. Add a small spatula measure of ascorbic acid crystals.

   e. Mix thoroughly and place in a hot water bath. Bring the mixture to a boil and then allow to cool.

   f. IF phosphate ions are present, a deep blue color will form due to the formation of the phosphomolybdate anion which reduces even more to form a mixed-valance complex, which gives the deep blue color.

   g. This ion reaction is written as

   $$PO_4^{3-}{}_{(aq)} + (NH_4)Mo_7O_{24}\cdot4H_2O_{(aq)} + C_6H_8O_{6(s)} \rightarrow [P(Mo_4)^{5+}(Mo_8)^{7+}O_{40}]^{7-}{}_{(aq)}$$

   (Deep blue color)

9. To test for *potassium ion,*

   a. Measure 13 mL soil solution into a test tube.

   b. Add 5 mL of 1% w/v sodium tetraphenylborate solution to the test tube.

   c. Mix thoroughly.

   d. IF potassium ions are present a white precipitate of potassium tetraphenylborate will form. In low potassium ion concentrations, the potassium tetraphenylborate may not precipitate out but form a colloidal solution, which will increase the turbidity or the solution.

   e. The reaction is written as $NaB(C_6H_5)_{4(aq)} + K^+_{(aq)} \rightarrow KB(C_6H_5)_{4(s)} + Na^+_{(aq)}$

10. Screen soil sample for *microorganisms*: Obtain petrifilm from your instructor; sterilize eyedropper or pipette with $\geq$70% Isopropyl Alcohol, then rinse with water sample twice before applying water to the petrifilm surface. Follow the instructions with the petrifilm for dispersal. Place in a safe, warm location to monitor growth over the next week, checking daily. Record results.

11. To test for pesticides, atrazine and/or simazine, use the test strips provided and follow the instructions associated with the strips. Record your results in the data table.

12. Clean and return your glassware, return or discard left over materials according to your instructor, and leave your lab area clean.

Data Table

Chemical Test	Results*	Further Observations**
pH		
Nitrogen		
Phosphorus		
Potassium		
Pesticide		

*Individual test kits will provide further information about the categories, depleted, deficient, adequate, and surplus/sufficient. We will use those for further determination of quantities of nutrients.

**Include anything you might have observed, color changes, amount of turbidity, etc.

VII. Conclusions

Answer the questions on the student report form (Appendix D).

VIII. References

1. Mosaic (2018). *Nitrogen*. Retrieved from http://www.cropnutrition.com/efu-nitrogen#nitrogen-in-plants.

2. Mosaic (2018). *Potassium*. Retrieved from http://www.cropnutrition.com/efu-potassium.

3. Mosaic (2018). *Phosphorus*. Retrieved from http://www.cropnutrition.com/efu-phosphorus.

4. Pioneer. (n.d.) *Growth Potential Corn Growers' Workshop*. Retrieved from http://www.pioneer.com/CMRoot/International/Australia_Intl/Publications/Corn_Workshop_Book.pdf.

5. United States Environmental Protection Agency. (2017 May 22). *Drinking Water Contaminants-Standards and Regulations*. Retrieved from https://www.epa.gov/dwstandardsregulations.

9. HERBICIDE BIOASSAY

I. Objectives

1. To use a bioassay to measure the dose/response of a test organism.
2. To determine the lettuce seed IC50 (concentration that causes 50% inhibition of germination) for selected herbicides.

II. Facts to Know

Bioassays and IC50s

All drugs, herbicides, pesticides, etc. are subjected to bioassays to prove their effectiveness. The bioassay determines the dose/response relationship for the test compound. The dose/response concept assumes that the test organism will respond to a given chemical compound in a predictable manner. At some upper threshold concentration, all of the test organisms will be killed (in our case, the seeds will not sprout). In the absence of the compound, all of the test organisms will live (in our case, all the seeds will sprout). The relative sensitivity of the test organism to the compound is characterized by the intermediate concentration that is capable of killing half of the test organisms. This is called the IC50, or inhibitory concentration 50%, or dose that inhibits 50% of the test organisms.

We will use seed germination because we will evaluate various herbicides. You would use this same assay if you were going to test whether pond or stream water contained herbicides, in which case it would be considered an environmental bioassay. In any case, the test organism that you choose for your bioassay must be sensitive to the types of compounds you are evaluating.

Each organism has its own set of sensitivities to compounds. Most plants are salt-intolerant and cannot grow in brackish water. For this reason, we will also carry out a reference toxicity test using NaCl. You will measure the germination response of your lettuce seedlings to a range of NaCl concentrations from 0 to 250 mM because we already know that this is the correct range.

Serial Dilutions

Before scientists begin an experiment, they search the scientific literature for papers related to their topic. That way, they design their experiments so that they are studying something that no one else has studied and so that they are doing the experiments in the same way as others. If you can't find a reference mentioning your compound, then you need to test a very wide range of concentrations and hope that you hit the right range. If your lowest concentration is too high, none of the seedlings will germinate and you would have identified an excellent herbicide. During the second
trial, you would repeat the experiment using a narrower range of concentrations.

Serial dilutions are the usual way to assess a broad range of concentrations. For example, 1 mg/L of a good herbicide should surely prevent all of the seedlings from sprouting. To cover a large range of diluted concentrations, you could start with a "10-fold dilution series." That is, you would test 1.0, 0.10, 0.010, 0.0010, and 0.00010 mg/L of the herbicide. Another common dilution series is the "2-fold dilution series." In this case, you would be testing 1.00, 0.500, 0.250, 0.125, and 0.0625 mg/L herbicide.

III. Community Connections
Herbicides and Human Toxicity

The following three herbicides are likely to be among the ones that you will test: paraquat, Dicamba, and glyphosate (sold as Roundup®). Each herbicide kills the plant in a different way and each is relatively nontoxic to humans. Humans don't have the enzymes inhibited by these pesticides. For instance, paraquat disrupts electron transport during plant photosynthesis by accepting the electron instead of allowing it to be passed to the usual biological acceptor. The paraquat radical then transfers the electron to any nearby molecular oxygen, which forms the very reactive superoxide radical. Superoxide radical reacts with all sorts of biomolecules in the plant. This degrades these plant's molecules, which prevents them from functioning properly. Since humans don't have the photosynthetic enzymes, paraquat is not nearly as harmful to humans as it is to the plants. Similarly, humans don't use auxins as hormones like plants and are not especially harmed by Dicamba. The least harmful of these herbicides is glyphosate because humans lack the shikimate pathway entirely.

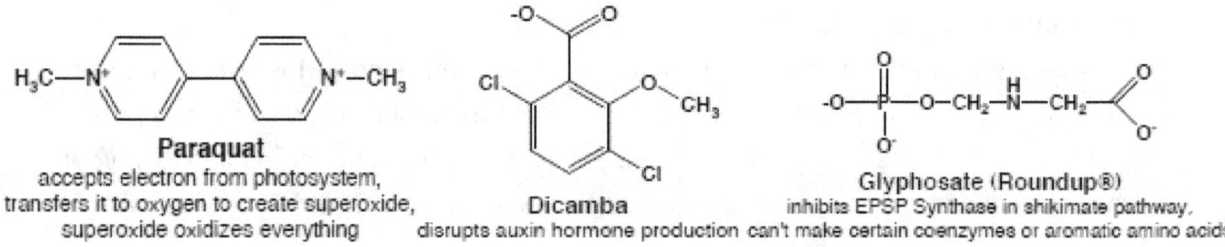

Paraquat
accepts electron from photosystem, transfers it to oxygen to create superoxide, superoxide oxidizes everything

Dicamba
disrupts auxin hormone production

Glyphosate (Roundup®)
inhibits EPSP Synthase in shikimate pathway, can't make certain coenzymes or aromatic amino acids

Until very recently, all commercial herbicides, drug candidates, etc. were tested in a mouse toxicity bioassay. This assay was performed to assess whether the compound was too toxic to use in humans. In the mouse toxicity assay, you determined the concentration of the substance that killed half of small population within a few days of ingestion. The resulting value is called the LD50, or lethal dose 50%, or dose capable of killing half the population. This bioassay is no longer performed because the concentration that causes death is much less important compared to the concentration that cures the disease *relative to* the concentration that causes humans to feel ill. Nevertheless, the following table of herbicide LD50s demonstrates that the herbicides you will test today are not very toxic to mice.

Herbicide Animal Toxicity Data	
Herbicide	**LD50 orally in rats, mg/kg**
Paraquat	100
Dicamba	1040
NaCl	3750
Glyphosate	4880

IV.   Safety Considerations and Materials

1. Always wear safety goggles;
2. Avoid skin contact with the herbicides by wearing gloves.

III.   Materials

Equipment *per student group*

Test tubes, 12, and test tube rack

Beakers, 2-50 mL

Thermometer

Weigh boats or petri dishes, 12

Filter papers, 7.5 cm diameter, 12

Marking pens

Graduated cylinder, 10 mL & 50 mL

Plastic wrap

Scissors

Graduated Pipette, 5 mL or eyedroppers

Samples and Reagents

Lettuce Seeds, $\geq$ 60

Household bleach, 10 mL

250 mM NaCl, 20 mL

1 mg/L herbicides, 5-mL of glyphosate, Dicamba*, or Paraquat

Distilled water

*Clarity Herbicide is 480 g/L plus inert ingredients

IV.   Procedure

We are thankful for the bounty of food that we are able to use lettuce seeds in this lab.

First Week

1. **Prepare the lettuce seeds.** Add 10 mL of household bleach to a clean beaker. You need a wide opening that will allow you to retrieve the seeds. Add at least 60 seeds to the bleach solution and let them sit for 20 minutes. Decant *(= pour liquid off keeping the solid in the beaker)* the solution into the sink but don't lose the seeds. Rinse and decant the seeds **five** times with distilled water. The bleach treatment kills any fungal spores that would interfere with seed germination.

You are washing the seeds with distilled water because tap water has unknown amounts of minerals and other compounds that could affect germination. That is, the tap water would introduce variability that would make it difficult to compare our results with those from other researchers using different water sources, or even the same source at different times of the year.

(You might have one partner do the NaCl dilution and the other do the herbicide dilution series in the interest of time)

2. **Prepare the NaCl standard solutions.** Bring *20 mL of 250 mM NaCl* to your desk in a **clean beaker**. Label a series of test tubes with the final NaCl concentrations listed in the table below. Then, add the *listed amounts of distilled water* and 250 mM NaCl to each tube using a *graduated pipette or eyedropper (using a 10 mL graduated cylinder determine how many drops = 0.5 mL and 1.0 mL)*

final [NaCl] (mM)	Water (mL)	250 mM NaCl (mL)
0.0 (Control)	5.0	0.0
25	4.5	0.5
50	4.0	1.0
75	3.5	1.5
100	3.0	2.0
250	0.0	5.0

3. Prepare **the Serially Diluted Herbicide solutions.** Choose one of the herbicides and record its name in your notebook. Label a series of test tubes with the final herbicide concentrations listed in the table below. Fill them with the listed amounts of distilled water. Add 5.0 mL 1 mg/L herbicide to the first tube. Transfer 0.5 mL of the first tube to the second tube. Swirl the contents to mix well. Continue the serial dilutions as indicated

Herbicide_____

final [herbicide] (mg/L)	Water (mL)	stock [herbicide] (mg/L)
1.0	0.0	5.0 mL
0.10	4.5	0.5 mL of previous solution

0.010	4.5	0.5 mL of previous solution
0.0010	4.5	0.5 mL of previous solution
0.00010	4.5	0.5 mL of previous solution
0.0	4.5	0.0

4. **Germination test.** You will need 12 weigh boats *or petri dishes*, one for each NaCl and herbicide solution that you prepared above. In the bottom of each weigh boat, place a piece of filter paper. Cut the paper so that it rests easily in the bottom of the weigh boat or *petri dish*. The paper is there to absorb the test solutions to keep the seeds moist so that they will germinate.

   We *can't* substitute paper towels or coffee filters for the paper filters until we prove they are nontoxic. Paper towels are often made of low-quality paper and contain polyphenols and iron, both of which can be toxic to plant seeds. Coffee filters may have retained some of the bleach that was used to make them white. Filter paper is bleached to make it white but it is then cleaned to very high standards because of its use in scientific experiments.

5. Add **2.0 mL** of each test solution to the weigh boats. Place five lettuce seeds onto each piece of wet filter paper and arrange them so that they do not touch each other or the sides of the weigh boat or *petri dish*. Seal the top of the weigh boat with plastic wrap to retain the moisture. Use a Sharpie marking pens to write the salt or herbicide concentration on the plastic wrap.

6. Incubate the seeds at constant temperature (use your thermometer to check the approximate temperature) until next lab period (7 days × 24 hours/day = 168 hours). Be sure to write in your notebook whether you have stored your samples in the *dark* (in your drawer). You are doing this to mimic the conditions of being in the soil. The amount of light the seeds receive during germination is an important factor for some herbicides but not for others. If two groups have chosen paraquat, then one group should grow them in the dark while another grows them in the light on the windowsill. Write your name on the weigh boats or *petri dishes* if you leave them on the windowsill.

7. Clean and return your glassware, return or discard left over materials according to your instructor, and leave your lab area clean.

Second Week

1. **Toxicity Data Analysis.** Count the number of seeds in each boat that have germinated. ***Draw a picture of one of your control seedlings and one of your partially inhibited seedlings this is on the last page of lab.*** Label the roots and shoot. Measure the root length of each seedling to the nearest millimeter. *You can also measure the shoot length if you wish but it is less sensitive to herbicides than root length.*

NaCl Standard Test Results

2. Germination Conditions (growth time, growth temperature, and light/dark):

Final [NaCl] (mM)	Number of Seedlings that Sprouted	Length of Roots (also calculate the average)	Other Observations
0.0			
25			
50			
75			
100			
250			

Herbicide Test Results

3. Germination Conditions (herbicide, growth time, growth temperature, and light/dark):

Final [Herbicide] (mg/L)	Number of Seedlings that Sprouted	Length of Roots (also calculate the average)	Other Observations
0.0			
0.00010			
0.0010			
0.010			

0.10			
1.0			

V.   Conclusions

Answer the questions on the student report form (Appendix D).

10. PLANT PIGMENTS: EXTRACTION, CHROMATOGRAPHY, & SPECTROMETRY

I. Objectives

1. To extract and identify chemicals contributing the colors of native dyes.
2. To learn the principles of spectrometry.
3. To determine the relationship between concentration and light absorption.

II. Facts to Know

We see colors because substances absorb different parts of the visible light spectrum (part of the electromagnetic radiation spectrum). In this laboratory experience, we will consider how things get their color from certain substances inside them. In this laboratory experience, the substances are food dye molecules.

The color that our eyes perceive is related to wavelengths of visible light (Table 10.1). It was Isaac Newton who, in 1672, divided the visible spectrum into the seven colors we remember today as the name ROY G BIV, where BIV is blue, indigo, and violet. The main change since then is that we now use the newer term "cyan" for the light blue or turquoise color between green and blue, and we use the word "blue" to describe dark blue things instead of the older word "indigo".

Table 10.1: Wavelengths of light associated with each color*

Color	Wavelength (nm)
Red	740 to 625
Orange	625 to 590
Yellow	590 to 565
Green	565 to 520
Cyan	520 to 500
Blue	500 to 435
Violet	450 to 380

*Wavelength ranges are from the *CRC Handbook of Fundamental Spectroscopic Correlation Charts*, edited by Thomas J. Bruno and Paris D. N. Svoronos, CRC Press, 2005.

A simple way to think about how a substance absorbs light is using a color wheel (Figure 10.1). If a substance absorbs a specific color, then the complementary color (directly across the wheel) is the observed color of that substance. For example, green leaves absorb light in the red region of the spectrum but not in the green. When we look at plants, see the green light that reflects off their surface or passes through them. Likewise, orange juice does not absorb light in the orange region but in the blue region. Substances that are colorless (or white) do not absorb light in any region of the visible spectrum, so our eyes perceive the substance as not having color. Substances that are very dark, or black, absorb light of many wavelengths.

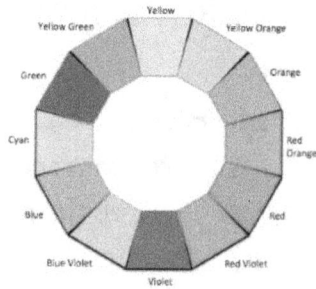

Figure 10.1: Color Wheel

Absorbance Spectrometer (or Spectrophotometer)

An absorbance spectrometer (Figure 10.2) measures how strongly a substance absorbs light at a given wavelength. The measurement is called absorbance, abbreviated as "A".

Spectrometers do this by detecting how much light passes through the substance. First, the beam of white light shines on a prism or diffraction grating to separate the light into its component wavelengths. A small portion of the light (centered on one wavelength) then passes through a sample inside in a holder called a cuvette. The cuvette is made of a substance that does not absorb light in the range you are measuring. Finally, the intensity of the light that emerges from the cuvette is measured by a light-sensing detector called a photomultiplier tube.

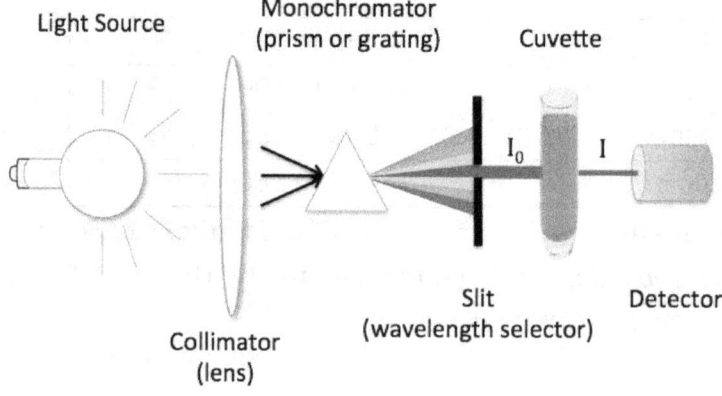

Figure 10.2: Absorbance Spectrometer Design

Quartz cuvettes are the most useful because they don't absorb visible light and only a little of ultraviolet light. Quartz cuvettes are expensive, however, and easy to break because quartz is very brittle.

Plastic cuvettes can be used for samples that absorb in the visible region of the spectrum because they don't absorb visible light. Plastic cuvettes cannot be used in the ultraviolet region of the spectrum because they are made of polystyrene, a substance that absorbs ultraviolet light. Instructions for your lab's spectrophotometer will be available in a separate handout.

Absorption Spectrum

An absorption spectrum is a plot of absorbance versus wavelength for a substance. Each substance's absorption spectrum is similar to a single fingerprint. You can use it to identify the class of substances in the cuvette but you can't use it to definitively identify the substance.

An example of an absorption spectrum is the one for iron (II) phenanthroline dissolved in water (Figure 10.3). Its spectrum begins at 350 nm in the violet region of the visible spectrum and ends at 700 nm in the red region. An examination of the iron phenanthroline spectrum shows that it absorbs weakly at 350 nm, begins increasing its absorption at about 365 nm, has what are called a "shoulders" at 440 nm and 475 nm, has a "peak" at 510 nm, and then decreases in absorption until it reaches a low at about 600 nm. The point of maximum absorption, 510 nm, is called the wavelength maximum (λ_{max}). Since this complex absorbs most between 425 and 550 nm, it is absorbing blue, cyan, and green light. From the color wheel, you would predict the color of the solution to be red orange, which it is.

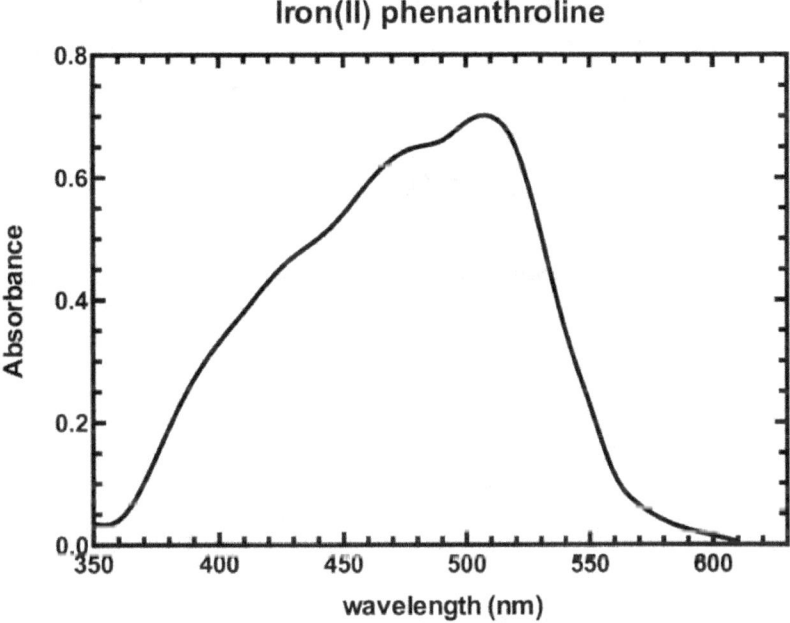

Figure 10.3: Absorption spectrum for the iron(II)phenanthroline complex.

Beer's Law: Using Spectrometry to Measure Concentration

One of the most useful features of spectrometry is that it can be used quantitatively to determine the concentration of a substance. This works because there is a direct relationship between the concentration of a substance, c, and the amount of light absorbed, A. This relationship is expressed mathematically in Beer's Law:

$$A = \varepsilon bc,$$

where ε is the molar absorptivity constant of the substance and b is the path length. The path length is the distance the light must pass through the solution. Most cuvettes are designed to have a 1-cm path length so that it remains a constant. Molar absorptivity is the proportionality constant. It is specific for each substance at each wavelength.

Most importantly, Beer's Law can be used to calculate concentration c after you measure absorbance A because the molar absorptivity for common substances is known. A typical method for determining the concentration of a substance is to measure Absorbance at the wavelength maximum (λ max) and a constant path length. The wavelength maximum is used because you get the strongest signal.

A typical experiment that demonstrates Beer's Law is when you prepare a series of samples of different concentrations, measure their absorbance, and then plot concentration versus absorbance. According to Beer's Law, the slope of the line is equal to εb. Here's an example for the iron (II) phenanthroline complex (Table 10.2) where you know the concentrations of your samples beforehand. The slope of the line in Table 10.2 indicates that the molar absorptivity of iron (II) phenanthroline complex is 14,680 M^{-1} cm.

Strategy for Dealing with Mixtures

Table 10.2: Absorption Standard Curve for iron (II) phenanthroline*

Concentration (M)	Absorbance
0	0
0.0000054	0.068
0.0000107	0.151
0.0000161	0.234
0.0000214	0.307
0.0000268	0.390

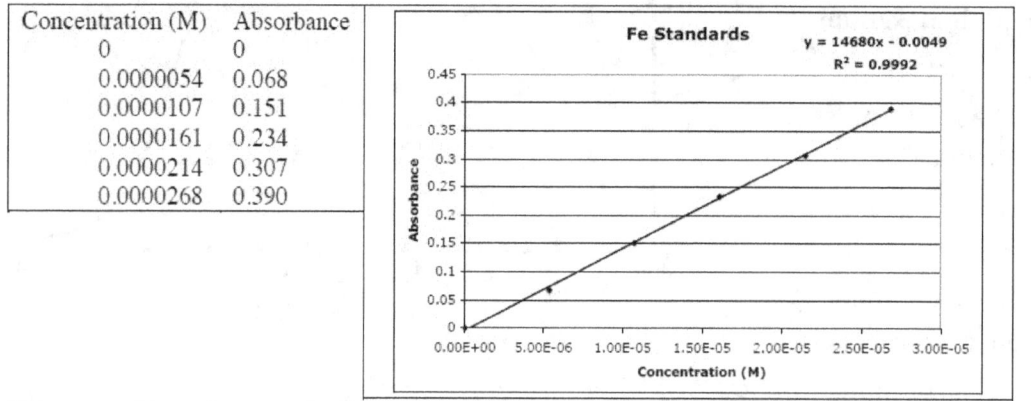

*These samples were prepared using an excess amount of phenanthroline and then adding different amounts of iron (II).

Absorbance is an additive property. In other words, if I have two substances in the same solution that both absorb light, then the absorbance of the solution will be the sum of their absorbances. Consider the case in which Compound A has a molar absorptivity of 5,000 $M^{-1}cm^{-1}$ at 600 nm and 50 $M^{-1}cm^{-1}$ at 400 nm and Compound B has a molar absorptivity of 20 $M^{-1}cm^{-1}$ at 600 nm and 20,000 $M^{-1}cm^{-1}$ at 400 nm. A mixture of these two compounds in solution would display the following absorptions if the path length were 1 cm.

At 600 nm... $A_{mixture} = A_{Compound\ A} + A_{Compound\ B}$

$A_{mixture} = \varepsilon_A b c_A + \varepsilon_B b c_B$

$A_{mixture} = (5000\ M^{-1}cm^{-1})(1\ cm)(c_A) + (20\ M^{-1}cm^{-1})(1\ cm)(c_B)$

At 400 nm... $A_{mixture} = A_{Compound\ A} + A_{Compound\ B}$

$A_{mixture} = \varepsilon_A b c_A + \varepsilon_B b c_B$

$A_{mixture} = (50\ M^{-1}cm^{-1})(1\ cm)(c_A) + (20,000\ M^{-1}cm^{-1})(1\ cm)(c_B)$

Ideally, when analyzing a solution containing a mixture of compounds, you would use a wavelength where the molar absorptivity of one compound is essentially zero while the molar absorptivity of the other compound is relatively high. Then, the light absorbed at this chosen wavelength is essentially due to only one compound. This is nearly the case in the above example. At 600 nm, 20 $M^{-1}cm^{-1}$ is very small compared to 5000 $M^{-1}cm^{-1}$ and we can assume that Compound B absorbs essentially no light at 600 nm. At 400 nm, 50 $M^{-1}cm^{-1}$ is very much smaller than 20,000 $M^{-1}cm^{-1}$ so Compound A absorbs essentially nothing at 400 nm. The previous equations then simplify to the following:

At 600 nm... $A_{mixture} = (5000\ M^{-1}cm^{-1})(1\ cm)(c_A)$

At 400 nm... $A_{mixture} = (20,000\ M^{-1}cm^{-1})(1\ cm)(c_B)$

Therefore, you would measure the concentration of substance A at 600 nm and Compound B at 400 nm even though there are two compounds in the cuvette.

III. Community Connections

Native plants provided more than food for Native American Indians. Plants were used for weaving baskets, utensils, housing, medicine, and the pigments were used in dying leather, artwork, and pottery and baskets.

Natural dyes were used for dying porcupine quills which were then used to decorate clothing, utensils, and jewelry. The color blue required extracts from these plants Larkspur, Beech, Wire Birch, and Indigo and the color green required Prince's Pine, Moosewood, and Evergreen (Prindle, 19914). Gilmore (1914) discussed the native plants uses by Great Plains

Indians along the Missouri and provided a table of plants and dyes, including their scientific name and names in native languages (Table 10.3).

To test natural pigments, natural fibers such as cotton, silk, or wool work best. The cloth needs to be treated with a mordant or fixative so that the dyes will adhere to the fiber and not easily wash out. Several common mordants include alum ($KAlSO_4$), iron ($FeSO_4$), and copper ($CuSO_4$). Alum is used in cooking and can be found in the spice section of grocery stores. The chemical name for alum is hydrated potassium aluminum sulfate, which you might have access to as well. Iron sulfate will work but darkens the dyes and by itself gives a brown color. Iron nails boiled in water works well also but will dull other dye colors. Copper sulfate is available commercially, but you can use copper pennies instead. However, using copper as a mordant will add a green hue to your dyes or can be used alone for green coloring. Common fixatives include salt (1 part NaCl to 16 parts water), tannins which occur naturally in some plants, vinegar (1 part CH_3COOH to 4 parts water), baking soda (0.5 cup $NaHCO_3$ to 1-gallon water), cream of tartar ($KC_4H_5O_6$) (often used with alum as the mordant), and washing soda (Na_2CO_3) (Sam, 2013).

Table 10.3. Native Plants and their pigment color (Gilmore, 1914). [a]

English	Dye Color	Scientific	Ho-Chunk	Lakota	Omaha & Ponca
Black walnut hulls	black	*Juglans nigra*	chak	hma	tdagë
Bloodroot	red	*Sanguinaria canadensis*	peh-hishuji		minigathe maka^n waü
Cottonwood buds	yellow	*Populus sargentii*		waga-cha^n	maa-zho^n
Dodder	orange	*Cuscuta lagenaria*	maka^n-chahiwicho		
Lamb's quarters	green	*Chenopodium album*		wahpe toto	
Lichens	yellow	*Usnea barbata*		cha^n-wiziye	
Smooth sumac	yellow	*Rhus glabra*	haz-ni-hu	cha^n-zi	mi^{n}bdi-hi
Soft maple twigs	black	*Acer saccharinum*	wissep-hu	tahado	wenu-shabethe-hi

[a] Superscript n is pronounced as "n" when your tongue touches the roof of your mouth.

IV. Safety Considerations

1. Wear safety goggles, wear gloves and lab coat to prevent staining your clothes and skin;
2. Do not look directly at the spectrometer as it uses an intense light beam.

V. **Materials**

Equipment	**Samples and Reagents**
Hot plate	Native plants, flowers, or fruit such as
knife	chokecherries, plums, black walnuts,
Blender	raspberries, blackberries, sunflowers
*Beakers, 400-600 mL	Distilled water**
Graduated cylinder, 10 mL &100 mL	
Test tubes, 6 medium or large & test tube	Chromatography paper (**white** *coffee filters,*
rack, grease pencil	*filter paper will also work)*
Centrifuge and tubes	Alum & cream of tartar solution***
Spectrophotometer and cuvettes	Muslin cloth, wool yarn ***
Funnel and filter paper	
Ring stand and round clamp	

* *Use plastic or glass for extracting and dying as the metals in metal containers react with some of the dyes, changing the colors and dying ability of the solutions.*

***Use only distilled water so ions in tap water do not interfere with chemicals from the plants*

****Only if going to use extracts for dying after absorption spectrophotometry. Many dyes need a mordant, which are usually an inorganic oxide that mixes with the dye to fix and stabilize the dye colors. Some plants contain chemicals that act as their own mordants. Common mordants are vinegar, alum, and wood ashes.*

VI. **Procedures**

Extraction of pigments

1. Gather your necessary lab equipment and materials, including your plant or fruit sample. The following are the extraction procedures for the samples.

    a. *Walnuts*: remover outer hull from 2 nuts, cut the hull into small pieces, and gently boil in about 180 mL water for about 10 minutes. Let solution sit for one hour and reheat to boiling. Keep solution at room temperature until ready to obtain an aliquot (*smaller amount from the sample*) for the spectrophotometer.

    b. *Blackberries, blueberries, chokecherries, & raspberries*: Mash 500 of fresh or frozen berries and pour off the juice. *(To use as a dye, heat juice to boiling, remove from heat, and add about $\frac{1}{8}$ tsp of alum.)*

    c. *Sunflowers*: Add 500 mL of petals, fresh or dried, to 240 mL water and boil gently for 3 minutes. Let simmer for another 15 minutes *(remove at least 20 mL before adding $\frac{1}{4}$ alum to use as a dye)* and store at room temperature until ready to obtain an aliquot for the spectrophotometer.

2. Filter your extract solution to remove the solid petals, fruit, or hull pieces. You might keep your filter paper as a reference for the color.

3. If you are still concerned about suspended particles (colloid) in your extract solution, centrifuge for 3 minutes. If there is no pellet of solid materials in the bottom of the centrifuge tube, you can skip this step. Decant carefully or use a transfer pipette to remove the liquid into a clean, dry beaker. You will need 10-30 mL of each extract.

Paper Chromatography Protocol

4. If you do not have chromatography paper, use a white coffee filter or regular filter paper. Cut into strips to fit into your test tubes or beakers. The length will depend on the height of the test tube or beaker you are using.

5. Set up at least two strips for each pigment extract, to compare water and an alcohol (Ethanol, preferably) as solvents. Draw a pencil line two cm from bottom of the strip. This is your start line. Use a capillary tube or toothpick to place your sample on the start line.

6. Pour your solvent (water into one and ethanol into the other) to just cover the bottom, at least less than 1 cm.

7. Put the strip of chromatography paper with samples into the test tube or beaker, so that the bottom just touches the solvent. ***Start line must stay above the solvent!***

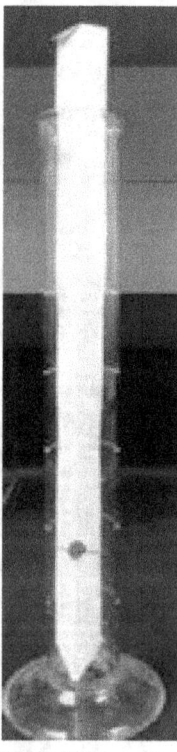

8. Allow the solvent to move ("climb") up the strip, "dragging" the soluble pigments with it.

9. Remove the paper from the chamber when the solvent front is 2 cm from the top or if the solvent front stops.

10. Carefully, mark the solvent front with a pencil and where the middle of the color spots is on the paper.

11. Measure the distance the solvent moved: from start to finish line; measure the distance from the start line to the color spots.

12. In the adjacent figure, you can see the final stage; initially you will place your extract above the solvent level in your containers; the solvent layer does not have to be deep, less than

13. Calculate the Rf of the solvent front and the middle of the color spot. Record in the table.
    RF equations:

    I.   Rf top $Rf_{(a)} = \dfrac{a}{d}$

    II.   Rf bottom $Rf_{(b)} = \dfrac{b}{d}$

    III.   Rf center $Rf_{(c)} = \dfrac{\left[\frac{a+b}{2}\right]}{d}$

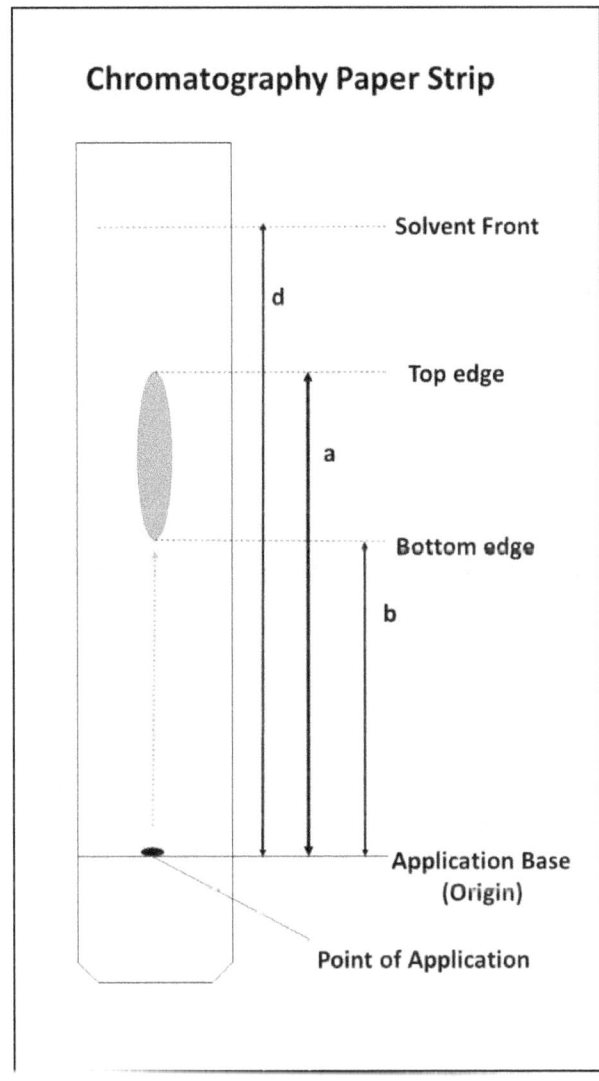

a. Water Solvent Data Table *(Use a separate row for each pigment that separates from one extract)*

Extract	Solvent front	Pigment front	R$_f$ Values

b. Alcohol Solvent Data Table *(Use a separate row for each pigment that separates from one extract)*

Extract	Solvent front	Pigment front	R$_f$ Values

Absorption Spectrometry

14. Follow the instructions for use of your spectrophotometer.

15. Label a series of test tubes with letters as below:

Tube Label	Extract, mL	Distilled Water, mL	Procedure	Dilution Factor
A	10	0.0	10 mL	Stock solution
B	5.0	5.0	5.0 mL previous solution	$\frac{1}{2}$
C	5.0	5.0	5.0 mL previous solution	$\frac{1}{4}$
D	5.0	5.0	5.0 mL previous solution	$\frac{1}{8}$
E	5.0	5.0	5.0 mL previous solution	$\frac{1}{16}$
F	0.0	10	0.0	Water control

16. Each cuvette has two sides that are smooth and clear, the light passes through these sides. Carefully wipe off those sides with a Kim-wipe to ensure any liquid and/or fingertips are removed.
17. Starting with the water control, fill the clean, dry cuvette; place in the spectrophotometer and take the absorbance spectrum. Scan over the visible light wavelengths. Save your scan, labeling it as the control.
18. Repeat with each dilution sample, working from most dilute to least dilute: water control, tube E, tube D, tube C, tube B, and last tube A. This reduces the chance of contaminating the more dilute samples. Rinse the cuvette with distilled water twice between each scan, drying the outside well.
19. After you have collected your spectra, clean up your lab area, covering your test tubes of sample if you wish to save them for next week. You can also save you stock solution (original extract solution) in a closed and labeled container.
20. Analyze your data. Identify maximum wavelength(s) for your extract.
21. If you were able to save your data in graph format great! If not, graph your data using Excel or other graphing program. Your TA/instructor will have instructions for you.

Pigments and Dyes

22. Prepare your cloth for dying by washing in plain water, no soap.
23. Soak your cloth in a mordant & fixative bath for 30 minutes. Wring gently excess solution but do not rinse or wash with water.
24. Immerse your cloth into a cool dye bath, then heat gently to simmer for approximately 30 minutes without stirring. Remove when you have the color intensity desired. Rinse to remove any excess dye, and then hang to dry.
25. To create designs on the fixed cloth, let it dry after removal from mordant & fixative bath to prevent the running of the dyes. Use an eye dropper or capillary tube to draw designs with the plant extract pigment. You may need to add more than one application of extract to achieve the desired darkness of dye. You might also experiment with direct dying using fruits and petals of the native plants.
26. Clean and return your glassware, return or discard left over materials according to your instructor, and leave your lab area clean.

VII.   Conclusions

Answer the questions on your student report form (Appendix D).

VIII.   References

1. American Chemical Society (2015). Chemistry Colors Our World. Web. Retrieved from https://www.acs.org/content/acs/en/education/outreach/celebrating-chemistry-editions.html.

2. *CRC Handbook of Fundamental Spectroscopic Correlation Charts*, edited by T.J. Bruno and P.D. N. Svoronos, CRC Press, 2005

3. Densmore, F. (1974). *How Indians Use Wild Plants for Food, Medicine & Crafts. (formerly called Uses of Plants by the Chippewa Indians).* Dover Publications, Inc. New York

4. Driessen, Kris. (n.d.). The Earliest Dyes. Quilt History. Retrieved from http://www.quilthistory.com/dye.htm.

5. Prindle, T. (1994). Natural Dyes for Porcupine Quills. Native Tech: Native American Technology and Art. Retrieved from www.nativetech.org.

6. Sam. (2013). Mordants and Fixatives. All Natural Dyeing. Retrieved from http://www.allnaturaldyeing.com/mordants-fixatives/.

7. Science Clarified. (2012). Dyes and Pigments. Advameg, Inc. Retrieved from http://www.scienceclarified.com/Di-El/Dyes-and-Pigments.html.

8. Turtle Mountain Community College. (2006). *RISE Experiments*.

11. ENDOTHERMIC AND EXOTHERMIC REACTIONS –HOT & COLD PACKS

I. Objectives

1. To observe endothermic and exothermic reactions related to common household chemicals.
2. To determine the ratio of chemicals that gives highest or lowest temperature measurement.
3. To demonstrate the concept of limiting and excessive reagents.

II. Facts to Know

How do you know if a chemical reaction has occurred? Several common observable characteristics of a chemical reaction are the production of a gas, the formation of a precipitate, the change of color or texture, or the transfer of energy. References provide background information with videos to assist you in understanding chemical reactions in addition to what has been and will be covered in class.

In the most common type of controlled chemical reactions, you mix the reagents so that they form a known product in the highest yield. While working out the ratio of reagents, it is useful to keep one reagent constant and to vary the other reagent. When one of the reagents is not used up we say the other reagent is "limiting" because it limits how much product is produced.

This lab will study the transfer of energy that occurs in chemical reactions. We will observe the exothermic and endothermic energy events, by mixing reagents that raise or lower the temperature, that is, we will make hot and cold packs. We will vary the amounts of the reactants to determine if there is an ideal ratio for the highest yield of heat or the lowest temperature.

Ammonium chloride, ammonium nitrate, calcium carbonate, and sodium carbonate are substances that have uses in and around the home. When these substances are dissolved in water, bonds are broken to form ions, energy is either released (hot pack) or absorbed (cold pack). Therefore, these are not true chemical reactions, but instead phase changes of the solids dissolving in liquids with energy either given off or absorbed.

III. Community Connections

Many common household items have medicinal uses. The ingredients for hot and cold packs may be purchased for other uses. For example, ammonium nitrate and ammonium chloride are common fertilizes for houseplants, gardens, and agricultural crops. Sodium carbonate, the active ingredient of washing soda, is useful for boosting dry laundry detergent's cleaning ability as well as removing a wide variety of organic stains. Calcium carbonate is the main ingredient in limestone, chalk, and egg shells, all different substances with their own unique characteristics. Calcium supplements to prevent and treat for osteoporosis may be a calcium carbonate base as well. Crushed egg shells can be added to your potted plants or garden soils as plants also require calcium. Calcium added to the soil around tomatoes seems to reduce blossom end rot and tip burn in cabbage. If you have these ingredients on hand, then you could also make your own a hot or cold pack.

Review the community topics (Appendix B) to identify possible community connections.

IV.   Safety Considerations

1. Wear safety goggles; flush with water will if you spill any of the chemicals, dry or in solution on your skin, gloves offer protection from skin irritation;
2. Be cautious touching the beakers as the temperatures will change to cold and warm.

V.   Materials

Equipment	Samples and Reagents
Beakers 150 mL, 4-5	Ammonium nitrate, NH_4NO_3, ≥ 50 g*
Digital balance	**OR** Ammonium chloride, NH_4Cl, ≥ 50 g
Spatula or spoon	Sodium carbonate, Na_2CO_3, (washing
Weigh boat or paper	soda), ≥ 100 g
Stirring rod	**OR** Calcium chloride, $CaCl_2$, ≥ 50 g
Thermometers**	Distilled water
Graduated cylinder, 50 mL	*Quantities are per group

**Probe ware works really well for this experiment.

VI.   Procedures

Cold Pack

1. Label four beakers, A, B, C, and D. Weigh out 5.0 g, 10.0 g, 15.0 g, and 20.0 g of NH_4NO_3 or NH_4Cl and place into the four beakers in order of weight, A, B, C, and D, respectively.
2. Measure and record in the table below the temperature of the water before you pour it in to each beaker, Initial measurement. *If you have a separate beaker (250 mL or larger) fill with at least 200 mL of water, measure its temperature.* Otherwise measure the water's temperature in the graduated cylinder before pouring into each of the beakers.
3. Place a thermometer into each beaker if you have enough, add 20 mL of water to each beaker and stir; observe the temperature, recording the final temperature. Do not stir with the thermometer, use a stirring rod. If you do not have a thermometer for each beaker, wait until the temperature has not changed for a minute, record, then proceed to the next beaker and repeat the above procedure with each beaker.

4. Record your data in the table below, circling the chemical you used.

Amount of NH_4NO_3 **or** NH_4Cl

Temperature, °C	Beaker$_A$	Beaker$_B$	Beaker$_C$	Beaker$_D$
Initial				
Final (Lowest Temperature Measure)				

5. **If there is more than one group**, or if you wish, repeat with the other chemical you did not use above.

Amount of NH_4NO_3 **or** NH_4Cl

Temperature, °C	Beaker$_A$	Beaker$_B$	Beaker$_C$	Beaker$_D$
Initial				
Final (Lowest Temperature Measure)				

Hot pack

6. Either use additional beakers, or clean and rinse well with distilled water the equipment from above. Use new weigh boats or paper.
7. Measure out *10.0, 15.0g, 20.0 g, and 25.0 g of CaCl$_2$*, placing each into the beakers labeled, A, B, C, and D, respectively. *IF* using *Na$_2$CO$_3$,* measure out *5.0 g, 10.0 g, 15.0 g, and 20.0 g,* and place into the beakers labeled A, B, C, and D, in order of weight. *Again, if you have a separate beaker (250 mL or larger) fill with at least 200 mL of water, measure its temperature.* Otherwise, measure the water's temperature in the graduated cylinder before pouring into each of the beakers
8. Record temperature of water before adding it to the chemical as above in step #3. Place a thermometer into each beaker if you have enough, add 20 mL of water to each beaker and stir; observe the temperature, recording the final temperature. Do not stir with the thermometer, use a stirring rod. If you do not have a thermometer for each beaker, wait until the temperature has not changed for a minute, record, then proceed to the next beaker and repeat the above procedure with each beaker.
9. Record your data in the table below, circling the chemical you used.

Temperature, ° C	**Amount of CaCl₂ or Na₂CO₃**			
	Beaker$_A$	Beaker$_B$	Beaker$_C$	Beaker$_D$
Initial				
Final (Lowest Temperature Measure)				

10. ___If there is more than one group___, or if you wish, repeat with the other chemical you did not use above.

Temperature, ° C	**Amount of CaCl₂ or Na₂CO₃**			
	Beaker$_A$	Beaker$_B$	Beaker$_C$	Beaker$_D$
Initial				
Final (Lowest Temperature Measure)				

11. Clean and return your glassware, return or discard left over materials according to your instructor, and leave your lab area clean.

VII. Conclusions

Answer questions on student report form (Appendix D).

VIII. References

1. Chemical Reactions. (2015 August). *Introduction*. Retrieved from https://sites.google.com/site/sfchemicalreactions/.

2. Chemical Reactions (2015 August). *Observable Characteristics*. Retrieved from https://sites.google.com/site/sfchemicalreactions/home/chemical-equations/symbols-and-numbers/reaction-types/general-forms/observable-characteristics.

3. Chemical Reactions (2018 April 3). *Reaction Types*. Retrieved from https://sites.google.com/site/sfchemicalreactions/home/chemical-equations/symbols-and-numbers/reaction-types.

12. MOLAR MASS OF BUTANE LIGHTERS

I. Objectives

1. To determine the density of butane
2. To determine the molar mass of butane

II. Facts to Know

In 1911, Dr. O. Walter Snelling at the US Bureau of Mines discovered how to compress propane, butane, and other hydrocarbons so that they were liquid at room temperature. This started the liquid propane industry that remains strong even today. In 1933, many manufacturers began producing liquid butane cigarette lighters after someone realized that the liquid butane was under enough pressure that its release could be easily controlled with a well-sealed valve. Butane is also so volatile that a flame can be started with just a few sparks. These lighters remain

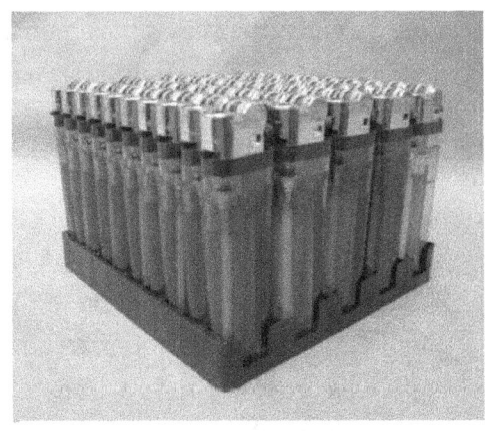

Package of Butane Lighters

popular because they don't require a flint or a wick or oil.

Butane is the name for any four-carbon hydrocarbon. There are two butane isomers: n-butane and isobutane. Isobutane has a lower boiling point than n-butane ($-$ 11.5 °C versus -0.5°C), which makes it more suitable for use at room temperature. The typical composition of the liquid in a butane lighter is 95% isobutane, 4% n-butane, and 1% of propane.

The ideal gas law, PV = nRT, relates pressure and volume with temperature and moles for most gases that aren't subjected to the extremes of pressure or temperature. In this experiment, the ideal gas law will be used to determine the moles of isobutane gas released from a butane lighter. You will also measure the mass of that released butane gas. The mass will be divided by the moles to yield the isobutane molar mass (or molecular weight). This will be compared to the molar mass that you calculate using its molecular formula of C_4H_{10}.

The volume of the gas will be measured by gas displacement just as you did in experiment 1 except on a larger scale in a cylinder. The pressure in the cylinder is a combination of both the pressure from the water vapor and the butane gas. Since the water pressure is determined by the atmospheric pressure, which changes with the weather, you will be given that day's barometric pressure to use in your calculation. The temperature of the gas can be assumed to be that of the water. The mass can be found by the difference in the mass of the lighter before the experiment and after the experiment.

III.   **Community Connection**

As described above, butane lighters have 1% propane in addition to the isomers of butane. Propane, C_3H_8, used in the home and for equipment is compressed and stored as a liquid. It is known as liquefied petroleum gas, LPG. LPG is used for space and water heaters as well as cooking in rural homes not serviced by a natural gas supplier (natural gas is methane, CH_4). LPG is also used to power small engines such as in mowers and irrigation pumps. Some vehicles and machinery use LPG for fuel.

Review the community topics (Appendix B) to identify possible community connections

IV.   **Safety Considerations**
1.   Always wear safety goggles;
2.   Liquid butane and butane gas are flammable; keep the lighter away from open flames.

V.   **Materials**

Equipment _per student group_

Graduated cylinder, 100 mL
Tub filled with room-temperature water
Thermometer
Towels
Electronic balance
Marking pens
Ring stand and clamp

Samples and Reagents

Butane lighter
Tap water, at room temperature

VI.   Procedures

1. Gather materials, put on your safety goggles.
2. Fill the tub with warm (less than 30° C) water about 2/3 full; open the butane lighter to the largest flame; immerse into the water; dry as thoroughly as possible. Weigh the lighter on balance, record the mass to the nearest 0.01 g in the data table.
3. Fill the 100-mL graduated cylinder with water and submerge it in your tub. Arrange the cylinder so it is upright with its mouth still underwater. There should be NO air bubbles in the cylinder. You can use a ring stand with clamp to hold the cylinder or hold it. (A partner could help!)

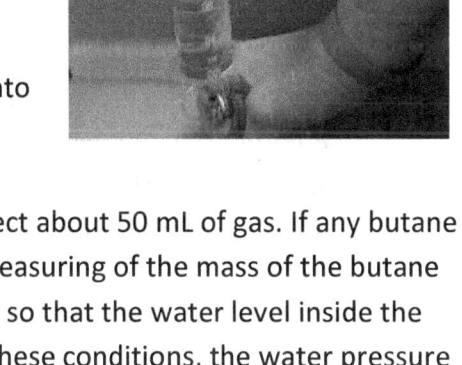

4. Record the temperature of the water bath. Convert the temperature ° C to Kelvin by adding 273.15 to the ° C measurement (°C + 273.15 = _____ Kelvin).
5. Holding the cylinder, lower the butane lighter into the water and put the opening of the lighter beneath the mouth of the cylinder. Press the lever on the lighter and collect about 50 mL of gas. If any butane bubbles escape, you must start over with the measuring of the mass of the butane lighter. **_KEEP_** the cylinder submerged but adjust so that the water level inside the cylinder is level with the water outside. Under these conditions, the water pressure is the same inside and outside the cylinder. Record the gas volume as this point.
6. Weigh the butane lighter after thoroughly drying to the nearest 0.01g. Subtract this value from the previous mass to calculate the mass of the liquid you collected as a gas.
7. Record today's barometric pressure (http://www.usairnet.com/weather/ or check the local weather forecast from another website). Conversion calculator for air pressure may be found at http://www.onlineconversion.com/pressure.htm or http://www.calculator.org/property.aspx?name=pressure.
8. Repeat these steps two more times.

Data Table

Butane Density Data					
Water Temperature _____ °C _____ K					
Today's Barometric Pressure _____ in Hg* convert to _____ atm					
	Volume (mL)	Mass (grams)			Density
Trial #		Butane Lighter Before	Butane Lighter After	Difference	(g/mL)
1					
2					
3					

9. According to the law of partial pressures, the total pressure of gas in the cylinder is the sum of the vapor pressure of water in the cylinder and the pressure of the butane.

10. To determine the pressure of the butane in the cylinder, you need to do the following calculation:

$$P(butane) = P(atmosphere) - P(water) = \text{_____} \; atm$$

Conversion equations: (or use the websites above)

Pressure in atm x 101,325 = pressure in pascal (Pa)

Pressure in mm Hg x 133.32 = pressure in pascal (Pa)

Pressure in torr x 133.32 = pressure in pascal (Pa)

Where P(atmosphere) is the barometric pressure and P(water) is the water vapor pressure from the table below:

Water Vapor Pressure			
Temp.	Pressure	Temp.	Pressure
(degrees C)	(atm)	(degrees C)	(atm)
0.0	0.0061	21.0	0.0246
5.0	0.0086	21.5	0.0253
10.0	0.0121	22.0	0.0261
15.0	0.0168	22.5	0.0268
15.5	0.0174	23.0	0.0278
16.0	0.0179	23.5	0.0286

16.5	0.0186	24.0	0.0295
17.0	0.0191	24.5	0.0304
17.5	0.0197	25.0	0.0313
18.0	0.0204	26.0	0.0332
18.5	0.0211	27.0	0.0351
19.0	0.0217	28.0	0.0372
19.5	0.0224	29.0	0.0395
20.0	0.0230	30.0	0.0418
20.5	0.0238		

Modified from http://kaffee.50webs.com/Science/activities/Chem/H2O.Vapor.Pressure.Data.html.

11. The molar mass is determined from the average mass of butane determined above divided by the moles of butane. The moles of gas molecules can be calculated from the information you have collected using the gas equation:

$$PV = nRT$$

where P is pressure in atm, V is volume in liters, n is in moles,
R is the gas constant $(\frac{0.08206 \; atm \cdot L}{mol \cdot K})$,
T is the temperature in Kelvin.

We will use PV= nRT to calculate the moles of butane, that is we will solve for n, which means the equation will look like $\frac{PV}{RT}$ = n.

12. Using the value calculated in #10, divide the gram butane by the moles butane to get the molar mass of butane. Calculate the average and standard deviation of butane's molar mass:

Trial #	Mass (g) butane from difference column	Divide by n	Molar Mass of butane
1			
2			
3			
			Average Molar Mass: Add the three Molar Mass values ÷ 3

Average Molar Mass + standard deviation (see Exps. 1 or 2):

_____ g/mol

13. Calculate the percentage error of your molar mass measurement & calculation with the known molar mass (58 g/mol):

$$\% \text{ error} = 100\% * \frac{MW theoretical - MW measured}{MW theoretical}$$

14. Clean and return your glassware, return or discard left over materials according to your instructor, and leave your lab area clean

VII.   **Conclusions**

1. Prepare your lab report using previous labs this semester as examples, as there is no student lab report form with this lab and subsequent labs. (You may copy the tables from the lab.)
2. What are two major sources of error in the experimental design?
3. If you were to do the experiment again, how would you change the experiment to reduce those two errors?
4. Give 1-2 examples where you would need to understand the molecular weight of a gas in the real world.
5. Summarize your results based on the lab objectives, your data, and an authentic application of the data *(5-10 sentences)*.

VIII.   **Instructions for lab report** (Appendix D)

Students will turn in a separate report, typed, that includes

a. Student name, date, and Campus in upper right corner
b. Title of Lab Centered
c. Objectives
d. Procedures: IF you followed procedures as given, simply state followed procedures. *IF* you changed procedures, or modified a step, include only those changes with an explanation why you modified the steps.
e. Data Table with results
f. Answers to questions in the lab and from *Conclusions*
g. Summarize your results based on the lab objectives, your data, and an authentic application of the data *(5-10 sentences)*.

13. FIRST SEMESTER CREATIVE PROJECT AND PRESENTATION

I. Objectives
1. To research a community topic of student interest.
2. To prepare a presentation, poster and oral, of results of research.

II. Facts to Know

Your learning is personal to you, and the content we talk about this semester will be meaningful to you in unpredictable ways. As we proceed through the semester, keep an eye out for topics and experiences that jump out to you, are related to the community topics listed in the Appendix B, and are related to chemistry. Think about how you might put your own spin on them for a creative project at the end of the course. You will have two lab periods to work on these with a third session for presentation your projects to the rest of the class.

Possible ideas for projects might be a short story, a film, a photo journal, a play, a costume, beadwork, a dance, a game, an experiment, a research paper … there are no limits as long as you can connect your project with chemistry! The creative project will be graded not only on the accuracy of the scientific information included, but also for originality and craftsmanship. A short proposal (roughly one paragraph) describing your plans for your project was due the week of midterms. Your lab instructor's approval was required for your project selection.

Your proposal should include the topic, the connection to chemistry, your chosen format for creating, and a list of the materials you think you need. If you do an experiment, you will have help with your lab design and collection of data. If you choose to do a research project, you will have help with research skills, paper format, and writing as well. You could do a research project and present in poster format instead of a research paper.

A rubric for your project with grading criteria will be provided once you have submitted your proposal and it has been approved by the lab instructor.

III. Community Connections

After selecting a topic of interest from the community issues (Appendix B), you will research that topic as related to chemistry concepts. You may choose to conduct an investigation/experiment. Your instructor will help you with experimental design, equipment and materials, and collecting data. You will have the remaining weeks of lab time to work on this project. Presentations will be during finals week, during the last lab session.

IV.   **Safety Considerations**

1. Depends upon your topic and research methods.

V.   **Materials**

Research topic from community issues related to chemistry

Poster-minimum size, 24" x 36"

Internet

Equipment and Materials if conducting an experiment as needed

Procedure for Preparing a Poster:

1. **How to create a poster for a science meeting**. There are three major things to consider when you present your research findings. The first is that you have full understanding of your hypotheses, methods, research findings, and interpretation. The second is the visual aspects of the poster itself. The third is the oral presentation you give to people who stop by your poster during the conference to learn about your research. The following description provides guidelines and tips for the second aspect, which is preparing a visually exciting poster that effectively communicates your findings. There are also tips about the oral presentation.

   a. **Poster Size**. Before you begin, you should consult the meeting guidelines about the poster size. You can use a wide range of programs for creating your poster but each of them requires you to set the image size before you begin adding content. You have been provided with a choice of two PowerPoint documents as a starting point. The "6 Poster Layout" is set to 40 x 46 and contains a mostly blank field. The "PosterPresentations.com" is set to 36 x 48 so you may have to adjust the size but it contains many fields just waiting for you to plug stuff into them. You can also search the internet to find many other poster templates. In fact, there is an insightful, relatively amusing, but overly long blog about creating posters that you might want to check out: http://colinpurrington.com/tips/academic/posterdesign.

      i. Overall, the poster should attract attention and convey information, use no jargon, have no spelling errors, use large text, be well-organized, and simple. The main sections of a poster are: title, abstract, introduction, results, conclusion, references, and acknowledgments. Here are suggestions for each of these sections.

   b. **Title.** The title area takes up the top section of the poster. It is common to place your school logo in the upper left and upper right corners. The title has three lines. The first line of the title should be in a very large font size (such as 100 points) that stretches across the top. The title should be short and avoid abbreviations. The second line gives the authors names in slightly smaller font size (such as 80 points). The presenter's name is listed first. The third line of the title gives the authors' institutional affiliations in an even smaller font (such as 60 points).

c. **Abstract**. Usually located in the upper left corner below the title field. Every meeting lists the number of words you can use in your abstract so you need to check that first. In fact, it is common to submit your abstract to the meeting, which is then used to determine whether you will be invited to give your poster. You are usually limited to 300 words or less to convince someone that your work is valuable. Therefore, the abstract has to state your purpose, goals, hypothesis, major finding, and possibly describe the major implications of your findings. Since the abstract is so short, you might consider writing it as follows:

   i. *Opening Sentence*: Describe an important issue to grab the reader's attention. This can be one of our Community Topics.

   ii. *Second Sentence*: Describe the long-range goal of your research area. A long-range goal addresses the issue you mentioned in the first sentence. For instance, describe a specific topic related to a Community Topic.

   iii. *Third Sentence*: Justify your long-range goal. Describe why your specific project is likely to yield results. State why you chose to study your specific topic.

   iv. *Next: State your hypothesis* (one sentence). State what instrumentation you used to carry out your research. Finally, describe the major findings.

   v. Strong closing sentence: Explain how your results address your hypothesis and bring you closer to reaching your long-term goal.

d. **Introduction**. Usually located below the abstract, in which case the abstract and Introduction take up the left-most fourth of your surface. The introduction describes the background that the listener would need to understand why you carried out your research. This would be a description of one of our Case Studies. It often includes a note about methods and instrumentation if they are special.

e. **Results**. Usually located in the middle section of the poster. This is the largest and most variable part of your poster. Use bold images with color highlights and font size 24 or larger. Make your graphs and tables first. As you add each new item to the poster, consider how you will use the images when describing your results at the meeting. The order of the images should help you describe the flow of ideas.

f. **Conclusions**. Usually located in the upper right corner below the title field. This is typically a short series of bullet points.

g. **References**. This is sometimes located at the end of the introduction unless you cite references in the Conclusions.

h. **Acknowledgments**. Usually in the lowest right-hand corner. List the funding source for the research and help of anyone who is not listed as a co-author.

2. **Tips for the oral presentation at a science meeting**. Every symposium chooses how long the oral presentations should be. For the purposes of this lab exercise we will choose 5 minutes. You will have only 1 min to describe the reason you did the research and 4 min to describe your research findings. Since everyone is best able to describe their research findings, this means you should focus on creating your 1-minute introduction and a very short take-home conclusion. Practice giving the oral several times out loud to yourself and several times to your friends. One last tip is to keep your finger or hand in place for several seconds when you are pointing to one of your images. It communicates a sense of purpose and calm to your listener as opposed to a jab in the general direction of the image which communicates you are in too great of a hurry.

Example of a Creative Project Grading Rubric

Grading Category	Points	Criteria
Connection	25%	25%: Elaborates significantly on chemistry & community topic 20%: Elaborates on chemistry & community topic 15%: Chemistry or connection lacking but not both 10%: Superficial chemistry connection 5%: Minimal chemistry connection 0%: No chemistry connection
Content	25%	25%: All chemistry information is accurate & current 20%: Most of the information is accurate & current 15%: Inaccuracies OR omissions 10%: Inaccuracies AND omissions 5%: Majority of information is inaccurate and incomplete 0%: Chemistry information is all inaccurate or missing totally
Creativity	25%	25%: Unique 20%: Best of similar projects 15%: Well done example of a category of products 10%: Average example of a category of projects 5%: Poor example 0%: Uncreative
Craftsmanship	25%	25%: Very well done, professional project 20%: Neat, well-done, effort is evident 15%: Nice looking but needs some corrections and refining 10%: Significant improvements to appear professional 5%: Little time & effort is evident 0%: No effort evident.

Second Semester Labs

14. ACID & BASE INDICATORS

I. Objectives
1. To sample different types of acid-base indicators
2. To compare chemical laboratory indicators with plant pigment indicators

II. Facts to Know

American Indians used indigenous plants for more than food. Plants are used for medicinal purposes, dying fabrics or other items, and art. We will use plant extracts for measuring the acidity of substances based on color changes.

We extracted pigments from native plants for chromatography and spectroscopy, particularly their fruits and nuts. In this lab, we will determine if a substance is an acid, base or neutral using a variety of plant extracts as well as known chemical indicators (litmus paper, pH or Hydrion paper, phenolphthalein or other dye indicators).

The word *acid* comes from Latin *acidus*, which means "sour or tart." It is also related to the Latin word *acetum,* which means "vinegar." Vinegar is a product of fermentation of apple cider or wines and has been known by early humans. Characteristics of acids include:
1. Sour taste;
2. Able to change color of litmus, a vegetable dye, from blue to red;
3. Able to react with
   a. Metals to produce hydrogen gas
   b. Hydroxide bases to produce water (H_2O) and an ionic compound known as a salt;
   c. Carbonates (limestone, chalk) to produce CO_2.

These properties are due to the presence of the hydrogen ion, H^+, released by acids in a water solution. Sometimes you will see hydronium ion instead, H_3O^+.

Related to acids, are bases, which are substances able to liberate hydroxide ions, OH^-, in a water solution. Water solutions of bases are called alkaline solutions or basic solutions. Characteristics of bases include:
1. Bitter or caustic taste;
2. Slippery, soapy feeling;
3. Able to change color of litmus from red to blue;
4. Able to interact with acids.

Acids, H^+, and bases, OH^-, may react to form water, HOH (H_2O), and a salt in a process call neutralization. For example, hydrochloric acid, HCl, may react with sodium hydroxide, NaOH, to form HOH and NaCl (salt).

III. Community Connections

Knowledge of acids, bases, and household substances may assist one's gardening skills and medicinal treatments. For example, in the soil quality testing lab, we tested the pH of the soil samples. Gardeners will want to treat their garden soils appropriately depending upon their intended crop. If one is growing blueberries, an acidified soil of 4.2-5.5 provides optimum growing conditions, whereas if you are growing peppers and tomatoes, preferred acidic growth conditions are 6.0 -6.8. Many common garden vegetables grow better soils with a pH between 6.0 – 7.5. A medicinal example involves Medicinally, one of the chemicals responsible for the sting of an ant bite or the sting of a bee or wasp is formic acid, CH_2O_2. Applying a solution of baking soda, ($NaHCO_3$ in water) will neutralize the acid, and thereby reducing the pain.

Review the community topics (Appendix B), to identify other possible community connections from your own history with this lab.

IV.   **Safety Considerations**

1. Always wear safety goggles.
2. Even dilute acids and bases may irritate skin; even if wearing gloves, wash hands before leaving lab.
3. Plant pigments may stain clothing and skin.

V.   **Materials**

Equipment	Samples and Reagents
Spot or well plates	Small (<20 mL) containers of acids, bases,
Litmus test paper strips	and neutral solutions including one (1)
pH or Hydrion test paper strips	unknown per group
Eye droppers or transfer pipettes	Phenolphthalein indicator solution
pH meter (optional)	Bromothymol blue indicator solution[t]
	Bromocresol purple indicator solution[t]
	Extracts of berries* and other aqueous
	pigments*

[t]**Additional indicators**-use what you have already if you do not have these.

*Suggestions: Berries-Chokecherries, cranberries, gooseberries, plums, raspberries, rosehips, strawberries; **Pigments**-Beets, black walnuts, cabbage, chamomile, mint, onion skins, purple petunia blossoms, spinach, sunflowers, willow bark

VI.   **Procedure**

1. Set up well plates
   a. Place a **few drops of acid** in the first three vertical wells (column) of a white porcelain or well plate.
   b. Place a **few drops of water** in the next three vertical wells (column)
   c. Place a **few drops of base** in the next three vertical wells (column)
   d. Place a **few drops of unknown solution** in the last three vertical wells (column).
2. Test each of the solutions of the first <u>horizontal </u>(row) with the paper strips of litmus paper and record the colors of the papers in the table. Repeat with Hydrion paper (if you have) and record the color in the data table. **Discard the papers in the waste basket!**
3. Add a **drop of phenolphthalein** to each of the four wells of the first horizontal row

and record the colors.

4. Add a **drop of Bromothymol blue[t]** to each of the next four wells of the second horizontal row and record the colors.
5. Add a **drop of bromocresol purple[t]** to each of the next four wells of the third horizontal row and record the colors.
6. Discard the contents of all wells into the sink, wash the wells with distilled water, and refill as before in step 1.
7. This time add enough of each "home-prepared" indicators to test each of the four solutions (acid, water, base, and unknown) and record the colors. (It may take more than several drops).
8. Clean and return your glassware, return or discard left over materials according to your instructor, and leave your lab area clean

Data Table

Indicator	Acid (HCl)	Neutral (Distilled H_2O)	Base (NaOH)	Unknown
Litmus				
Hydrion paper				
Phenolphthalein				
Bromothymol Blue				
Bromocresol Purple				
Petunias				
Cabbage juice				
Beet juice				
Tea				
Willow bark				
Spinach				
Cranberries				
Onion Skins				

If you need more rows, please add them.

What was the label on your unknown vial? _____

Was your unknown an acid, a base, or neutral? (circle your answer)

VII.   **Conclusions**

1. What radical (charged ion) is found in all acids?
2. What radical (charged ion) is found in bases?
3. What is the pH scale? Copy and label a pH scale into your report. Reference your source for the information.
4. What classification, acid, base, or neutral, is most of the food we consume? Again cite your source(s) if you use more than the data from this lab.

VIII.   **Instructions for lab report** (Appendix D)

Students will turn in a separate report, typed, that includes

a. Student name, date, and Campus in upper right corner
b. Title of Lab Centered
c. Objectives
d. Procedures: IF you followed procedures as given, simply state followed procedures. *IF* you changed procedures, or modified a step, include only those changes with an explanation why you modified the steps.
e. Data Table with results
f. Answers to questions in the lab and from **Conclusions**
g. Summarize your results based on the lab objectives, your data, and an authentic application of the data *(5-10 sentences)*.

15. ASCORBIC CONTENT IN TRADITIONAL NATIVE FOODS

I. **Objectives**

1. To determine the concentration of ascorbic acid, vitamin C, in foods.
2. To titrate with acid/base indicators.
3. To identify native foods ascorbic acid content.

II. **Facts to Know**

Vitamin C is a water soluble vitamin. Its chemical name of ascorbic acid tells us that it has anti-scorbatic properties, which means it prevents and cures scurvy. Although it is important for good health, humans do not have the ability to make their own vitamin C and must obtain it through diet or take it in vitamin supplements. Citrus fruits, potatoes and some green vegetables are known to be good sources of vitamin C (Table 14.1). Plants synthesize the compound for the growth, development, and protection of the plant. The exact pathway for its synthesis is not well understood.

Table 15.1. Amount of Vitamin C in Various Plants.

Plant source	Amount of Vitamin C (mg /100g)	Plant source	Amount of Vitamin C (mg /100g)
Rose hip	426	Spinach	30
Red pepper	190	Potato	20
Parsley	130	Green Beans	16
Broccoli	90	Tomato	10
Brussels sprouts	80	Watermelon	10
Elderberry	60	Banana	9
Strawberry	60	Carrot	9
Orange	50	Apple	6
Cantaloupe	40	Lettuce	4
Grapefruit	30	Raisin	2

USDA National Nutrient Database for Standard Reference Release 28, October 20, 2015, https://ods.od.nih.gov/pubs/usdandb/VitaminC-Content.pdf

Orange juice may lose half of its vitamin C in a week in the refrigerator. The vitamin C in cut fruit lasts longer (Some examplesof vitamin C loss after six days: mango, strawberry, and watermelon: less than 5%, Pineapple: 10%, Kiwi: 12%, and Cantaloupe: 25%, after 6 days). Cooking also destroys vitamin C (Table 14.2).

Table 15.2. Freezing fresh vegetables requires blanching, i.e., boiling for a short period to stop enzyme actions in the plant.[a]

Amount of vitamin C, mg/100g			
Vegetable	Raw	Frozen	Frozen then
Green beans	12.2	9.7	4.1
Broccoli	89.2	64.9	40.1
Peppers, sweet red	190.3	55.8	55.6

[a]USDA National Nutrient Database for Standard Reference Release 28, October 20, 2015 https://ods.od.nih.gov/pubs/usdandb/VitaminC-Content.pdf.

Vitamin C is required for the synthesis of collagen, an important structural component of blood vessels, tendons, ligaments and bone. It also is important in the synthesis of the neurotransmitter norepinephrine, which is critical to brain function and can affect mood.

Vitamin C is also a highly effective antioxidant. In small amounts, vitamin C can protect indispensable molecules in the body, such as proteins, lipids (fats), carbohydrates, and nucleic acids (DNA and RNA), from damage by free radicals and reactive oxygen species that can be generated during normal metabolism as well as through exposure to toxins and pollutants. Some recommended dietary allowances for vitamin C are listed in Table 14.3.

Table 15.3. Recommended Dietary Allowances (RDAs) for Vitamin C.[a]

Age	Males (mg/day)	Female (mg/day))	Pregnancy (mg/day)	Lactation (mg/day)
0–6 months	40*	40*		
7–12 months	50*	50*		
1–3 years	15	15		
4–8 years	25	25		
9–13 years	45	45		
14–18 years	75	65	80	115
19+ years	90	75	85	120

*Adequate intake; Smokers require 35 mg/day more vitamin C than nonsmokers.

[a]Oregon State University, Linus Pauling Institute, http://lpi.oregonstate.edu/mic/vitamins/vitamin-C.

As the name indicates, vitamin C, that is ascorbic acid, is an acid. Therefore, a base such as sodium hydroxide can neutralize it. Ascorbic acid is also easily oxidized. Both of these reactions can be used for quantitative analysis of the compound. However, there are additional acids in juices, fruits, and foods which may interfere with obtaining good results with the acid-base titration. We use the oxidation of the vitamin C in foods by iodine to determine the vitamin C's concentration When ascorbic acid reacts with iodine, the ascorbic acid is oxidized (loses electrons) and the iodine is reduced (gains electrons). Reduced iodine cannot react with starch. When all the ascorbic acid has reacted, any added iodine will then be able to react with the starch. A purple-blue-black color will form (and remain), giving you the end-point for the titration.

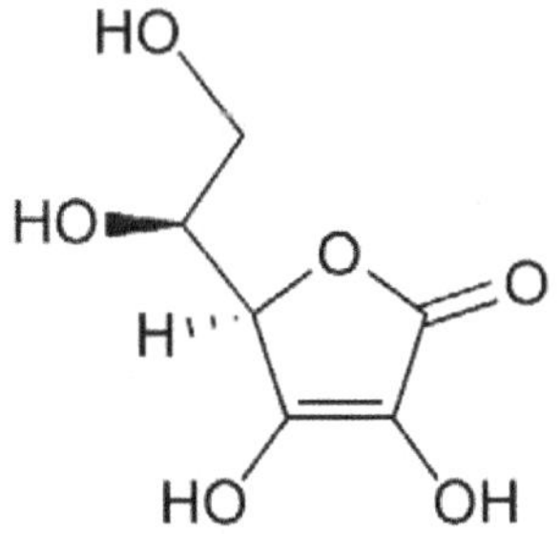

Figure 15.1. The structure of ascorbic acid, $C_6H_8H_6$ (MW = 176.12 g/mol).

The reactions are as follows:

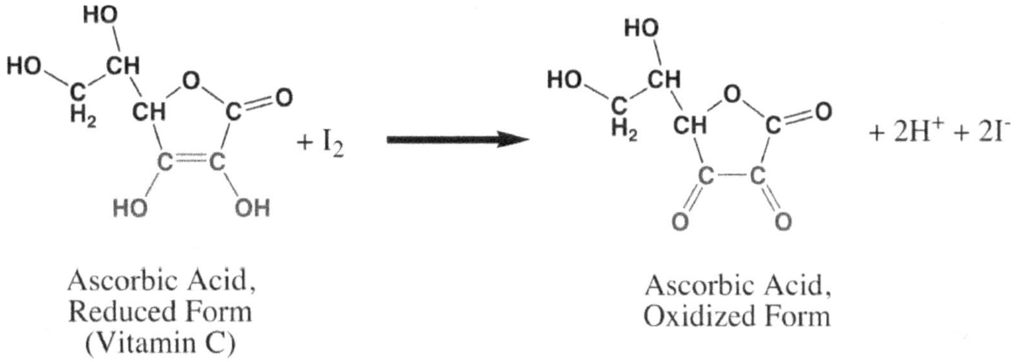

Ascorbic Acid,
Reduced Form
(Vitamin C)

Ascorbic Acid,
Oxidized Form

$$I^- + I_2 \longrightarrow I_3^-$$
(purple)

$$I_3^- + starch \longrightarrow starch\ complex$$
(purple)   *(blue-black)*

III.   **Community Connections**

There is a concerning trend in the health of Americans related to nutritional related diseases. In particular, there is an increase in Type II Diabetes (adult-onset) that may be addressed by eating more traditional foods, native plants, and more fruits and vegetables.

Fruits and vegetables are excellent sources of many different vitamins and essential amino acids, including ascorbic acid. Food quality depends on the ability to grow quality food.

Benefits of ascorbic acid and recommended doses for ascorbic acid can be found at the following websites:

- National Institutes of Health (https://ods.od.nih.gov/factsheets/VitaminC-HealthProfessional/)
- U.S. National Library of Health (https://medlineplus.gov/druginfo/meds/a682583.html)
- Live science (https://www.livescience.com/51827-vitamin-c.html)

Review the Native case study, When Our Water Returns: The Gila River Indian Community and Diabetes (http://nativecases.evergreen.edu/collection/cases/when-our-water-returns).

IV. **Safety Considerations**
1. Wear eye protection at all times.
2. Iodine stains both skin and clothing, wear gloves and lab coat/aprons.

V. **Materials**

Equipment	Samples and Reagents
Burette, 50 mL	Vitamin C standard solution, (1.0 mg/mL)
Funnels & filter paper, ring stand with ring clamp	1% Starch solution
Erlenmeyer flask, 125 mL	Acidified Iodine solution
Graduated cylinders, 10 & 100 mL	Various fruits & vegetables, $\geq$ 100 g each
Knife	(Preferably pink, red or dark green)
Mortars & pestles/Blender	Distilled water
Sand, for grinding in mortar, not blender	
Magnetic stirrer & stir bar (optional)	

VI. **Procedures**

Preparation of food extract

We are thankful for the bounty of food that we are able to use in this lab.

1. Chop about 100 g sample of a fruit or vegetable, cover with distilled water to grind in mortar & pestle or add 50 mL of distilled water in a blender.
2. Strain the mixture, washing the filter with distilled water.

3. Add distilled water to make a final solution of 100 mL of fruit/vegetable extract.

Standardizing Solutions

4. Add 25.0 mL of vitamin C standard solution ($\frac{1.0 \text{ mg}}{\text{ml}}$) to a 125 mL Erlenmeyer flask.
5. Add 10 drops of 1% starch solution. Swirl to mix well.
6. Rinse your burette with a small volume of the iodine solution and then fill it. Record the initial volume in Date Table 1.
7. Titrate the standard solution, remembering to swirl as you add the iodine solution to the flask. As you add the iodine solution slowly, you will see a purple/ blue-black color at point of contact. When the color is slow to disappear, proceed drop-by-by drop. When the entire solution turns blue-black and remains blue-black, stop. This is the end-point of the titration. Record the final volume of the iodine solution.
8. Repeat two more times. Ensure you have thoroughly cleaned the flask between each titration, or use three different clean flasks.

Data Table 1: Vitamin C Standardization Titration

Trial	Initial Burette Volume (mL)	Final Burette Volume (mL)	Iodine solution used (mL)
1			
2			
3			
Average & SD			

Titration of food extract

9. Add 25.0 mL of food extract to a clean 125 mL Erlenmeyer flask.
10. Add 10 drops of 1% starch solution. Swirl to mix well.
11. Fill your burette as needed. Record the initial volume of iodine solution in Data Table 2.
12. Titrate the standard solution, remembering to swirl as you add the iodine solution to the flask. As you add the iodine solution slowly, you will see a purple/ blue-black color at point of contact. When the color is slow to disappear, proceed drop-by-by drop. When the entire solution turns blue-black and remains blue-black, stop. This is the end-point of the titration. Record the final volume of the iodine solution.
13. Repeat the titration two more times. Ensure you have thoroughly cleaned the flask between each titration, or use three different clean flasks.

Data Table 2: *Fruit or vegetable used_____*

Trial	Initial Burette Volume (mL)	Final Burette Volume (mL)	Iodine solution used (mL)
1			
2			
3			
Average & SD			

14. Clean and return your glassware, return or discard left over materials according to your instructor, and leave your lab area clean.

Titration Calculations

15. Calculate the average amount of iodine solution used with the standard vitamin C solution and your food extract. Record in your data tables.

16. We will use the following proportion to determine the amount of vitamin C in your fruit juice sample. Remember, your standard vitamin C solution has a concentration of **1.0** $\frac{mg}{mL}$ *and we used 25.0 mL of our standardized vitamin C solution.*
    The proportion is:

$$\frac{\text{average value of iodine used for standard solution (mL)}}{25.0 \text{ mg}} = \frac{\text{average value of iodine used for fruit juice (mL)}}{X \text{ mg}}$$

Rearrange the above equation to solve for the unknown vitamin C content:

$$X \text{ mg} = \frac{(\text{average iodine used for fruit juice (mL)}) (25.0 \text{ mg})}{\text{average iodine used for standard solution (mL)}}$$

We used 25.0 mL of food extract so to find out the number of mg/mL, divide your answer by 25 mL, for $\frac{mg}{mL}$ vitamin C in your food extract.

VII.   Conclusions

1. Does the color of the food sample play a role in the titration results?
2. How is the concentration of the ascorbic acid affected by the level of grinding/blending?
3. Why is this a quantitative test rather than a qualitative test for ascorbic acid?
4. Why does the solution stay blue-black at the endpoint?
5. What was the concentration of vitamin C in your food extract? Compare it to given samples in the background information or in the USDA table

   (United States Department of Agriculture. (2015). *USDA National Nutrient Database for Standard Reference*. Retrieved from https://ods.od.nih.gov/pubs/usdandb/VitaminC-Content.pdf.)

6. Using the data provided in the background information, what is the daily recommended vitamin C requirement for a 10-year old child? How many servings of food extract would meet the vitamin C requirement for that child?
7. Summarize your results based on the lab objectives, your data, and an authentic application of the data *(5-10 sentences)*.

VIII.   References

1. Brown, J. (2009). When Our Water Returns: Gila River Indian Community and Diabetes. Evergreen State University. Retrieved from http://nativecases.evergreen.edu/collection/cases/when-our-water-returns.

2. College of Science. (n.d.). Determination of Vitamin C Concentration by Titration. University of Canterbury. New Zealand. Retrieved from http://www.canterbury.ac.nz/media/documents/science-outreach/vitaminc_iodine.pdf.

3. Katz, D. (2013). Determination of Vitamin C in Foods. Chymist.com Retrieved from http://www.chymist.com/Determination%20of%20Vitamin%20C%20in%20Foods.pdf

4. Helmenstine, A. M. (2018). Vitamin C Determination by Iodine Titration. ThoughtCo. Retrieved from https://www.thoughtco.com/vitamin-c-determination-by-iodine-titration-606322.

16. QUALITATIVE TESTS FOR ALCOHOLS

I. **Objectives**

    1. To qualitatively test for primary, secondary, and tertiary alcohols Lucas Reagent Test.

    2. To identify an unknown using these tests.

II. **Facts to Know**

Alcohols, aldehydes, and ketones are organic molecules that have recognizable oxygen-containing functional groups. These functional groups react differently with various reagents. An unknown structure may be identified as an alcohol, aldehyde, or ketone by performing a set of reactions and noting the results.

Alcohols are a class of organic compounds that contain the -OH functional group covalently bonded to a carbon atom. An alcohol may be represented as a water molecule in which one of the hydrogens is replaced with a hydrocarbon R group. Like water, alcohols are polar; however, their polar character decreases as the size of the nonpolar R group increases.

Alcohols can be further classified as primary, secondary, or tertiary. These terms describe the carbon atom that is attached to the –OH group.

primary   secondary   tertiary

The carbon atom in primary alcohols has one R group and two hydrogen atoms bonded to it. Secondary alcohols have two R groups and one hydrogen atom. Tertiary alcohols have three R groups. In the above examples, R is a methyl group, but any aliphatic or aromatic group may be bonded to the hydroxyl carbon. All three classes of alcohols are capable of hydrogen-bonding in the pure state or to other alcohol molecules because the -OH

105

functional group is always present. Alcohols also have the ability to hydrogen bond with water; however, their solubility decreases as the size of the R group and their hydrocarbon-like character increases.

Aldehydes and ketones are classes of organic compounds in which oxygen has a double bond to a carbon atom. The carbon and oxygen are collectively referred to as a carbonyl group. Aldehydes have one R group and one hydrogen attached to the carbon. Ketones have two R groups.

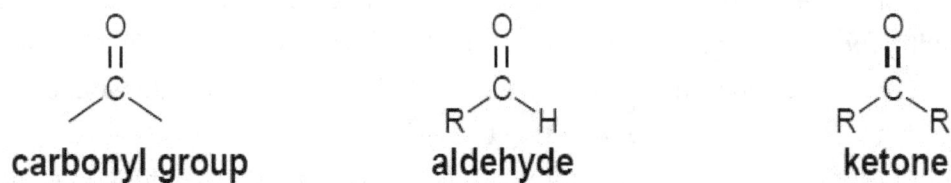

The geometry around the carbon atom of a carbonyl group is different from the tetrahedral geometry of sp3- hybridized carbon atoms attached to -OH functional groups. The presence of the double bond results in sp2 hybridization on the carbon and a trigonal planar arrangement of atoms.

Oxidation of Alcohols

Alcohols undergo several types of chemical reactions. The most important reaction alcohols undergo is oxidation to carbonyl compounds. Primary alcohols are oxidized to aldehydes, which themselves are sensitive to oxidation to carboxylic acids. Oxidizing agents convert secondary alcohols to ketones, and tertiary alcohols lack the ability to undergo oxidation.

The Lucas Test

The Lucas Test is used to identify whether a hydroxyl group, −OH, is present, which means it is used to identify alcohols. Although not all alcohols. Alcohols may be grouped by the number of carbon atoms attached to the same carbon the hydroxyl group is attached. Three classifications of alcohol are primary, secondary, and tertiary. Primary alcohols have only one carbon atom is attached this carbon, secondary have two, and tertiary have three.

$$\text{HO} - \underset{\underset{\text{H}}{|}}{\overset{\overset{\text{H}}{|}}{\text{C}}}\text{CH}_3$$

Ethanol, primary alcohol, 1°

$$\text{HO} - \underset{\underset{\text{CH}_3}{|}}{\overset{\overset{\text{H}}{|}}{\text{C}}}\text{CH}_3$$

2-Propanol. secondary alcohol, 2°
(isopropyl alcohol)

$$\text{HO} - \underset{\underset{\text{CH}_3}{|}}{\overset{\overset{\text{CH}_3}{|}}{\text{C}}}\text{CH}_3$$

2-Methyl-2-Propanol, tertiary alcohol, 3°
(t-butyl alcohol)

The reagent, called Lucas reagent, used in the Lucas test is a mixture of hydrochloric acid and zinc chloride. The hydroxyl group in alcohols may be replaced by the Cl from the hydrochloric acid when the catalyst, zinc chloride, is present. Not all alcohols readily react with the Lucas reagent.

As the alcohols react to form the alkyl halides, an intermediate cation is formed, called carbocation. All three classes of alcohols may form this carbonium ion or carbocation.

Tertiary alcohols react with the Lucas reagent immediately to form a cloudy suspension. Secondary alcohols react more slowly, producing the cloudy suspension within 3-5 minutes. Primary alcohols do not react under room temperature conditions, if at all. Heating in a water bath, some primary alcohols may react with the Lucas reagent.

Tertiary alcohols form the most stable intermediate and the stable intermediates form faster than unstable ones. The order of intermediate stability is 3°>2°>1°, which is why the tertiary react quickly and first.

Phenols do not react with the Lucas reagent either. If missed with it, a cloudiness may result from the insolubility of the alkyl halide in the reaction mixture.

The reaction equations of 1°,2°, and 3° alcohols are

$$R-CH_2OH \ + \ HCl \ \xrightarrow{\ ZnCl_2\ } \ R-CH_2Cl \ + \ H_2O$$

1° alcohol $\qquad\qquad\qquad\qquad$ clear after 5 minutes

$$R-\overset{\displaystyle R}{\underset{}{\overset{|}{C}HOH}} \ + \ HCl \ \xrightarrow{\ ZnCl_2\ } \ R-\overset{\displaystyle R}{\underset{}{\overset{|}{C}HCl}} \ + \ H_2O$$

2° alcohol $\qquad\qquad\qquad\qquad$ cloudy within 3-5 minutes

$$R-\overset{\displaystyle R}{\underset{\displaystyle R}{\overset{|}{\underset{|}{C}}OH}} \ + \ HCl \ \xrightarrow{\ ZnCl_2\ } \ R-\overset{\displaystyle R}{\underset{\displaystyle R}{\overset{|}{\underset{|}{C}}Cl}} \ + \ H_2O$$

3° alcohol $\qquad\qquad\qquad\qquad$ cloudy within 1 minute

III. Community Connections

Alcohol is found in a variety of food and over-the-counter medications. For example, many food extract flavorings contain alcohol such as almond (32%), pure peppermint (89%), and pure vanilla (41%). Cough medicine for night use can contain up to 25% alcohol. Knowing what the alcohol content of these items and other foods may be important if your children accidently ingest them.

Review the community topics (Appendix B) to identify possible community connections.

IV. Safety Considerations

1. Always wear eye protection.
2. Lucas reagent contains concentrated hydrochloric acid, HCl, that will vigorously react with human tissue. IF you get any on you, wash it off immediately with cool water (continue flushing for 15 minutes) and inform your instructor.

V. Materials

Equipment	Samples and Reagents
Test Tubes, 5-10 small, 7.5 cm, and rack	Lucas Reagent:
Test tube holder, individual	Concentrated HCl and $ZnCl_2$
Transfer pipettes or eye droppers	Variety of Alcohols
Boiling water bath:	1-Butanol or ethanol or 1-propanol
Hot plate	2-butanol, 2-pentanol, or 2-heptanone
Thermometer	Tertiary-Butanol or 2-methylpropan-2-ol
Beaker, 250 or 400 mL	Hand sanitizer
Hot mitt/holder	Unknown alcohol sample
Lab coats and gloves	

VI. Procedures

1. Set up the hot water bath in the hood. Place the 250 mL beaker half full of water on a hot plate and bring to a boil.
2. Meanwhile, label your test tubes and add 5 drops of each alcohol sample into separate clean, small test tubes.
3. Add 10 drops of Lucas Reagent to each sample and mix the contents well by agitation.
4. Wait one (1) minute, observing continuously, after mixing. Watch for a precipitate to form (cloudiness) to form.
5. Place the test tubes that are still clear into the boiling water bath. Observe the contents after two minutes in the boiling water bath.
6. Record your observations in the data table below.
7. Dispose of the test tube contents according to your instructor's directions.
8. Clean and return your glassware, return or discard left over materials according to your instructor, and leave your lab area clean.

Data Table: Record response with Lucas Reagent using + to mean positive and — to mean negative

Compound	Room Temperature	Boiling water Bath

VII.   **Conclusions**

1. What did your unknown test as, primary, 1°, secondary, 2°, tertiary, 3°, or not alcohol?
2. What type of alcohol, primary, 1°, secondary, 2°, tertiary, 3°, is in the hand sanitizer?
3. Would you ever use this test at your home? Why or why not?
4. Summarize your results based on the lab objectives, your data, and an authentic application of the data *(5-10 sentences)*.

17. QUALITATIVE TESTS FOR ALDEHYDES AND KETONES

I.   Objectives
1.   To qualitatively test for aldehydes and ketones
2.   To observe the properties and characteristic reactions of aldehydes and ketones
3.   To identify an unknown using these tests.

II.   Facts to Know

Molecules with ketones have a distinct aroma and are used in a plethora of scented products like candles, incense, and smudge sticks. These ketone molecules are key ingredients from the essential oils of plants.

Aldehydes and ketones are common in nature. They are known for their strong odors, some sweet and others pungent. Several examples of aldehydes and ketones from plants include vanillin from vanilla beans, cinnamaldehyde from cinnamon bark, 2-heptanon from blue cheese, and carvone which is from both spearmint and caraway seed. The musk smell of deer and other animals is from ketone molecules. Hormones are examples of ketones such as cortisone, progesterone, and testosterone. The major similarity between an aldehyde and a ketone is the carbonyl group. A carbonyl group is a carbon atom doubly bonded to an oxygen atom, as depicted below. Both molecules have a carbonyl group, the aldehyde and ketone differ in what atom is bonded to the carbonyl carbon. An aldehyde is bonded to one hydrogen atom and one R- group. A ketone is bonded to two carbons. Remember that the 'R' symbolizes any carbon side-chain, from one carbon to a million carbons. What it comes down to is that in an aldehyde the carbonyl group is on the terminal (last) carbon and the ketone's carbonyl group is not. Below are a few examples of aldehydes and ketones.

Formaldehyde   Acetone   Methyl Ethyl Ketone   Cinnamaldehyde
(MEK)

Three test reagents will be used in this experiment: Tollens' reagent, Iodine reagent, and Benedict's solution.

Tollens' reagent is used to differentiate between aldehydes and ketones. Aldehydes may be oxidized relatively easily, while ketones are not easily oxidized. The Tollens' reagent, composed of a silver-ammonium ion, will have its silver reduced by an aldehyde. This silver will plate a glass surface creating a mirror.

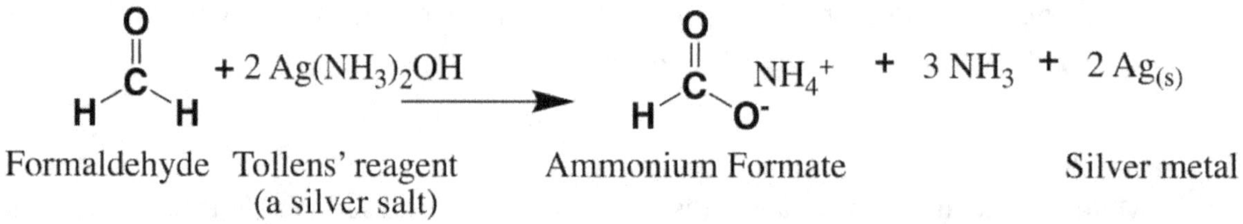
+ 2 Ag(NH3)2OH
NH4+
+ 3 NH3 + 2 Ag(s)
Formaldehyde
Tollens' reagent (a silver salt)
Ammonium Formate
Silver metal

Iodine^{3-} reagent (I2 + KI) reacts with acetaldehyde and all methyl ketones to form a yellow iodoform. Iodoform, CHI_3 is a yellow solid with a strong medicinal smell. Primary and secondary alcohols because they are easily oxidized to acetaldehydes and methyl ketones give a positive iodoform test. Iodoform will precipitate out of a mixture of methyl ketone or acetaldehyde, iodine and base.

Methyl Ketone

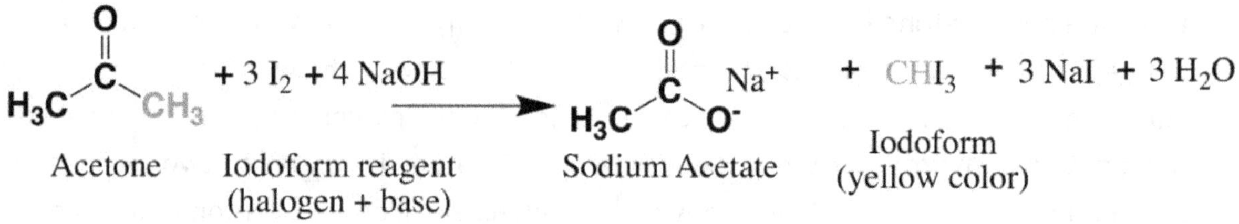
+ 3 I2 + 4 NaOH
Na+
+ CHI3 + 3 NaI + 3 H2O
Acetone
Iodoform reagent (halogen + base)
Sodium Acetate
Iodoform (yellow color)

Benedict's reagent also oxidizes aldehydes. Benedict's reagent is a basic solution of copper (II) citrate (whose complex formula cannot be represented by a simple formula). Benedict's reagent is a clear blue solution due to the Cu^{2+} ion but when it reactions with an acetaldehyde is converts to the insoluble, brick red Cu^{1+} ion.

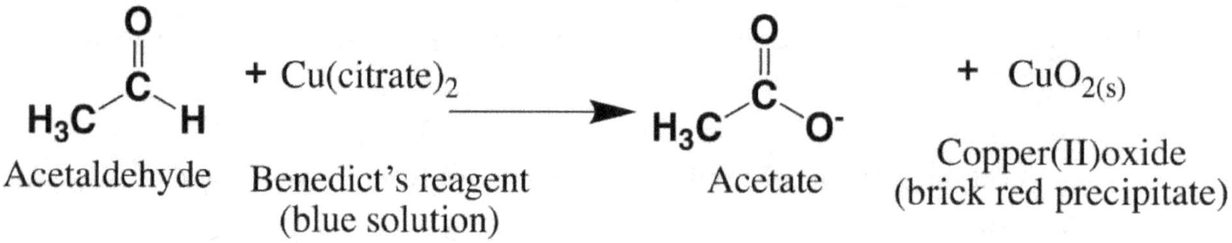
+ Cu(citrate)2
+ CuO2(s)
Acetaldehyde
Benedict's reagent (blue solution)
Acetate
Copper(II)oxide (brick red precipitate)

III. Community Connections

The plant genus, *Salvia* L., or sage, is a common source of natural fragrance. It also has culinary and medicinal uses. In North America, a variety of *Salvia* plants, native and introduced, were used by Native American Indians. *S. apiana*, or white sage is used in ceremonies and is sacred to a number of Tribal nations. White sage leaves were used as a cold remedy by either eating the leaves, smudging in a sweat house, or even smoked. Other species of sage found in North America were used for analgesics as a wash or poultice, as a salve for sores, as a decoction of leaves for colds, stomach ailments, even epilepsy, either as a wash, smoke, or a tea. *Salvia* species include *S. dorrii*, grayball and purple sage, *S. lyrata*, lyreleaf sage, and *S. mellifera*, black sage.

Some Salvia spp. are native to North America while others such as *S. officinalis*, also known as common, kitchen, or garden sage, was introduced from European travelers. Pliny the Elder (CE 23-79) is credited as the first one to describe in writing how *Salvia* was used by the Romans, which was most likely *S. officinalis*. One of the active chemicals in this and other sages is thujone, which is a chemical formula of $C_{10}H_{16}O$.

Thujone is a ketone with a menthol odor. It is best known as one of the compounds in the spirit absinthe, albeit in smaller quantities than originally report in the early 20th Century. In the United States, the Food and Drug Administration actually regulates the amount of thujone that can be found in food and other products. For example, thujone may be up to 50% of what is found in sage and sage oil, but foods or beverages must be thujone-free.

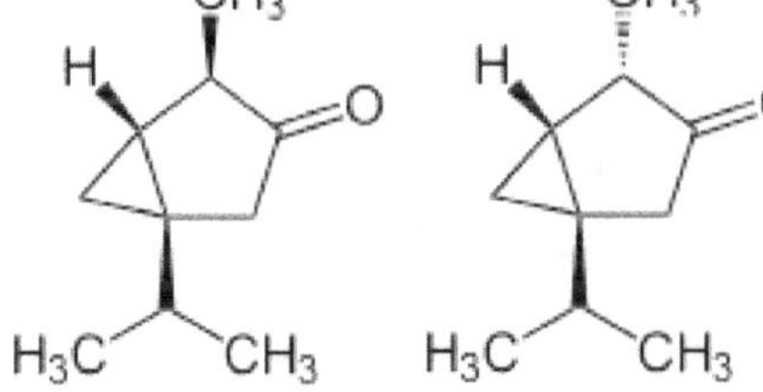

Figure 17.1. Structure of (−)-α Thujone (left), and (+)-β Thujone (right).

Review the community topics (Appendix B) to identify possible community connections.

IV. Safety Considerations:

1. Always wear eye protection.
2. Do not let the Tollens' reagent stand around, since it may form explosive substances; clean up those test tubes as soon as results are evident.
3. Dispose of reagents in the appropriately labeled waste beakers in the hood.

V. Materials

Equipment	Samples and Reagents
Test tubes, 12 small, 7.5 cm, and rack	Distilled H_2O
Test tube holder, individual	Tollens Reagent: 6 M Sodium
Pipettes, transfer or eye droppers	Hydroxide (NaOH), 0.1 M Silver Nitrate
Beakers, 50 ml, 250 or 400 mL	($AgNO_3$), & 6 M Ammonia (NH_3 or
Graduated cylinder, 10 ml	NH_4OH)
Boiling water bath:	Iodide^{3-} Reagent: (I_2—KI) solution
Hot plate	10% NaOH
Thermometer	Benedict's reagent*
Hot mitt/holder	Variety of Aldehydes and Ketones:
Lab Coats and Gloves	Acetone, Formaldehyde, & Methyl
	Ethyl ketone
	Unknown aldehyde or ketone samples

*Benedict's reagent is a solution made by dissolving hydrated copper sulfate, sodium citrate, and anhydrous sodium carbonate.

VI. Procedures

Solubility Test:

1. Gather three clean test tubes and label Sol 1, Sol 2, & Sol 3.
2. Place 2 mL of distilled H_2O in each of the three test tubes, then add
   a. To the first, add 10 drops of acetone.
   b. To the second, add 10 drops of formaldehyde.
   c. To the third, add 10 drops of methyl ethyl ketone.
3. Gently shake each test tube, and note the degree of solubility on the data table.

Tollens' Test:

4. Gather three clean 10-cm test tubes and label Tollens 1, Tollens 2, and Tollens 3.
5. Put 15 drops of dilute (6M) sodium hydroxide (NaOH) into each test tube and allow them to stand for at least 1 minute. This prepares the glass surface for the Tollens' test.
6. After 1 minute, empty the NaOH from the tubes and, without rinsing the tubes, add 15 drops of 0.1 M silver nitrate ($AgNO_3$) solution.
7. Next, add 15 drops of dilute (6M) ammonia (NH_3 or NH_4OH), shaking until the precipitate that forms initially dissolves. You have now created fresh Tollens reagent- $Ag(NH_3)_2$.
8. Add 5 drops of acetone to tube 1, 5 drops of formaldehyde to tube 2, and 5 drops of Methyl Ethyl Ketone to tube 3.
9. Shake by swirling the test tubes to mix and set the tubes aside. Observe them after 3

minutes. Set them aside for 7 more minutes. A silver mirror will develop if the test is positive. If no precipitate forms and no mirror appear it may be too cool; heat the tubes in a water bath to confirm that no reaction will occur.

10. Record your observations in Data Table 2.

    CAUTION: Do not let the Tollens' reagent stand around, since it may form explosive substances. Dispose of it in the appropriately labeled waste beaker in the hood. Destroying the Tollens' reagent: Add a few drops of HNO_3 to the Tollens' reagent after you are done with the test.

Iodoform Test:

11. Gather three clean 10-cm test tubes and label Iodo 1, Iodo 2, and Iodo 3.

12. Place 10 drops of acetone to tube 1, 10 drops of formaldehyde to tube 2, and 10 drops of Methyl Ethyl Ketone to tube 3.

13. Next, add 20 drops of I_2—KI solution to each tube.

14. Add 3 drops of 10% NaOH to each tube and shake by swirling to mix.

15. If a light yellow cloud does not form immediately after adding NaOH, swirl more vigorously for a minimum of 1 minute. A positive test results in a yellow cloudiness to form, replacing the original brown color of the Iodine[3-] reagent. A negative result means the original brown color remains, there is no cloudiness, although sometimes a brown layer will form.

16. Record your observations in Data Table 3.

Benedict's Test:

17. Gather three clean test tube and label Ben 1, Ben 2, and Ben 3.

18. Place 2 mL of Benedict's solution in each of the three test tubes.

    a. To the first, add 20 drops of acetone.

    b. To the second, add 20 drops of formaldehyde.

    c. To the third, add 20 drops of methyl ethyl ketone.

19. Heat the test tubes in the water bath provided for 10 minutes (in the hood). Watch for the appearance of reddish Cu_2O.

20. Record your observations on each tube Data Table 4.

21. Dispose of the test tube contents according to your instructor's directions.

22. Clean and return your glassware, return or discard left over materials according to your instructor, and leave your lab area clean

Data Table 1: Solubility Test

Sample	Tube Number	Observations
Acetone		
Formaldehyde		
Methyl Ethyl Ketone		

Data Table 2: Tollens' Test

Sample	Tube Number	Observations
Acetone		
Formaldehyde		
Methyl Ethyl Ketone		

Data Table 3: Iodoform Test

Sample	Tube Number	Observations
Acetone		
Formaldehyde		
Methyl Ethyl Ketone		

Data Table 4: Benedict's Test

Sample	Tube Number	Observations
Acetone		
Formaldehyde		
Methyl Ethyl Ketone		

VII. Conclusions

1. Include data tables in your lab report.
2. How do you account for differences in solubility between the three compounds tested?
3. **Tollens' Test**. Which compound(s) gave a positive test? Why?
4. **Iodoform Test**. Which compound(s) gave a positive test? Why?
5. **Benedict's Test**. Which compound(s) gave a positive test? Why?
6. Summarize your results based on the lab objectives, your data, and an authentic application of the data *(5-10 sentences)*.

VIII. References

1. Moerman, D.E. (1990). *Native American Ethnobotany.* Timber Press. Portland, Oregon.

2. Wikipedia. (2018 February 17). *Salvia*. Retrieved April 2018 from https://en.wikipedia.org/wiki/Salvia.

3. Wikipedia. (2018 March 29). *Thujone*. Retrieved from https://en.wikipedia.org/wiki/Thujone.

18. THE EFFECT OF ALCOHOL ON BETACYNIN EXTRACTION

I. Objectives

1. To investigate the effects of an alcohol on a biological membrane.
2. To identify betacyanin in the plant extract.

II. Facts to Know

The deep red color of garden beets may be used for food coloring and perhaps even textile dying. That deep red color is from the pigment, betacyanin. Betacyanin (Figure 17.1) is one of two pigments found in the betalain family of alkaloid pigments found in some families of plants. Betalains are found in plants belong to the order Caryophyllales and some fungi (e.g. Fly agaric). Plants that have betalains do not have the more common pigment, anthocyanin. There are two types of betalains found in these plants: betacyanin which is red colored and betaxanthus which is yellow in color. Both of these pigments give color to flowers, fruits, and roots. Betalains are found in the vacuole of the plant cells and are water soluble.

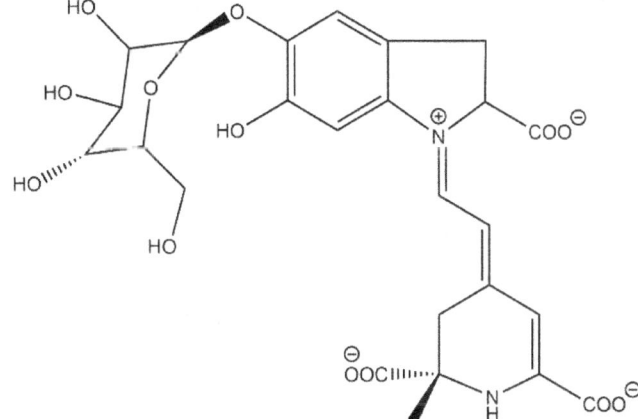

Figure 18.1 Betacyanin

Edible beets are *Beta vulgaris* which includes many varieties created over time by selective breeding. One variety provides the red tuber we eat or use for food coloring. Another variety is the sugar beet which has been bred to contain around 17% sugar. Sugar beet leaves may provide feed for cattle and sheep although there is a variety that the tubers are harvested for livestock fodder as well. The leaf beet varieties are harvested for human consumption and the plants include chard and spinach beet.

The betacyanin are kept inside the vacuole by the membrane. Membranes are composed mainly of lipids and proteins. If membranes are damaged, they can no longer perform their function of controlling what moves in and out of cells or in this case the vacuole inside the cell. Several factors may damage cell membranes such as changes in pH and temperature. Organic compounds such as alcohols and detergents can damage or destroy the membrane as well.

Ethanol, C_2H_5OH (Figure 17.2), is the type of alcohol that is present in alcoholic beverages (beer, wine and all other types of liquor). Ethanol is a polar molecule due to its methyl end and alcohol end. Ingestion of ethanol has been done since the beginning

119

of recorded history. Remember from our solid density lab using chocolate, the pulp of the cacao fruit was used to make fermented beverage? Originally, ethanol was believed to have medicinal properties. For example, many cough syrups have ethanol added to help dissolve the active ingredients. Ethanol may also be added as a preservative (Zorn, 2017).

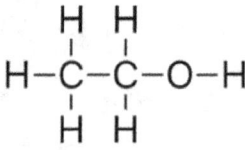

Figure 18.2. Ethanol

III. Community Connections

Originally, beet greens were consumed similarly to chard, which is a variety of *Beta vulgaris*. The thin, fibrous roots were not eaten by early humans although they were occasionally used in medicine. The bulbous root shape we are familiar with and do eat began to emerge in the last 1500s in Germany or Italy (Avery, 2014).

Both tryptophan and betaine are found in beets and are associated with a sense of well-being. Therefore, they were consumed for healthy well-being. In addition, red beets contain iron, and often are eaten to supplement one's iron intake. However, there is only about 1 mg of iron in approximately a cup of red beets.

Review the community topics (Appendix B) to identify possible community connections.

IV. Safety Considerations

1. Always wear safety goggles.
2. Beet juice will stain clothing and hands, wear gloves and a lab coat.

V. Materials

Equipment	Samples and Reagents
Beaker, 400 mL	Detergent
Graduated cylinder, 10-mL	Ethanol
Hot plate	Two large beets
Knife and Fruit corer	Distilled water
Pipettes	Lint free tissues
Test tubes, 11 medium size, & rack	
Spectrophotometer & cuvettes	
Stop watch/timer	

VI. Procedures

We are thankful for the bounty of food that we are able to use in this lab.

Finding wavelength for beet color using absorbance

1. Prepare a solution of 5 mL of ethanol and 5 mL of distilled water in a test tube, fill a cuvette with this solution as your blank.

2. Place a small piece of beet in the test tube with ethanol-water mixture.
3. Allow the beet to stay in the solution until it is a dark, pink color, around 30 minutes.
4. Pipette the liquid into a cuvette.
5. Find the absorbance spectrum for the pigment, betacyanin, measure the absorbance for the beet solution at 10 nm intervals between 400 and 550 nm. Zero your spectrophotometer using your blank. The highest absorbance reading corresponds to the wavelength for the betacyanin that you will need for the rest of this lab.

Data Table

Wavelength (nm)	Absorbance	Wavelength (nm)	Absorbance
400		480	
410		490	
420		500	
430		510	
440		520	
450		530	
460		540	
470		550	

Maximum absorbance occurs at _____ nm.

Beet Preparation
6. Insert a corer into a beet that has been peeled and sliced.
7. Slice the beet cores into equal portions.
8. Wash the pieces of beet in cold water until no more pigment washes out. (Cover the pieces of beet with a damp towel if you are not going to use immediately.) There should be no ragged edges; no outer skin on any pieces; all about the same size; need to fit in the test tubes without being wedged, poked, or squeezed into the tube.

Alcohol Solution Preparations
9. Obtain 11 clean test tubes, labeled 1 through 11.
10. Add the following amounts of water and ethanol to each tube:

*Wavelength from earlier in this lab

Test Tube	Water, mL	Alcohol, mL	Concentration of Alcohol (%)	Absorbance at _____ nm*
1	10	0	0	
2	9	1		
3	8	2		
4	7	3		
5	6	4		
6	5	5		
7	4	6		
8	3	7		
9	2	8		
10	1	9		
11	0	10	100	

11. Calculate the % concentration of ethanol in each tube and finish filling in the table above and below.

Effects of Ethanol on Beet Membrane

12. Place a piece of beet in each test tube.
13. Gently shake the tubes every 2 minutes, for 10 minutes.
14. After 10 minutes, pipette at least 5 mL of solution from each test tube into a cuvette, starting with the lowest concentration of alcohol.
15. Measure the absorbance *at the wavelength we found in the first part of this lab*. Zero your spectrophotometer with distilled water.
16. Record the absorbance in the table above.
17. Clean and return your glassware, return or discard left over materials according to your instructor, and leave your lab area clean.

VII.   Conclusions

1. Provide your data in table or graph format and attach to lab, concentration (x-axis) versus absorbance (y-axis).
2. What was the maximum absorbance of betacyanin?
3. Based on your results, at what concentration is the most damage done to the beet membrane, supporting your answer with your results.
4. Why did the beet pieces need to be the same size?
5. Why did we rinse the beet pieces until no more pigment came out?
6. What shouldn't we squeeze, poke, or wedge the beet into the test tube?
7. Explain your data: was there a steady increase in the damage to the cell membranes as concentration increase? What is a possible explanation for this?
8. Explain how the alcohol damages the cell membrane.
9. How is measuring the absorbance of the betacyanin concentration an indication of damage to the cell membrane?
10. Summarize your results based on the lab objectives, your data, and an authentic application of the data *(5-10 sentences)*.

VIII.   References

1. Avery, T. (2014 October 8). Discover the History of Beets. Retrieved from http://www.pbs.org/food/the-history-kitchen/history-beets/.

2. Missouri Botanical Garden (2018 April 23). *Beta vulgaris*. Retrieved from http://www.missouribotanicalgarden.org/PlantFinder/PlantFinderDetails.aspx?kempercode=a668.

3. Science & Plants for Schools. (2018 April 23). Using beetroot in the lab. Retrieved from http://www.saps.org.uk/secondary/teaching-resources/754-using-beetroot-in-the-lab.

4. Turtle Mountain Community College. (2006). RISE Experiments.

5. Zorn, P. (2017). What Cold Medicines Contain Alcohol? *Healthfully*. Retrieved March 2018 from https://healthfully.com/cold-medicines-contain-alcohol-5554043.html.

19. PREPARATION AND IDENTIFICATION OF ESTERS

I. Objectives

1. To prepare an ester from an organic acid and an alcohol.
2. To prepare, name, and identify the scent of a variety of esters

II. Facts to Know

What is an Ester?

Esters are associated with aromas and flavors. Esters occur naturally and can be derived from a variety of plants, since these natural esters change during the ripening process, synthetic esters are used as food additives to enhance flavor.

Other esters are used as industrial solvents and are the starting materials for plastics. Fats and oils are also esters. They are composed of long-chain "fatty" acids.

How is an Ester formed?

Carboxylic acids react with alcohols to form compounds known as esters (R represents any hydrocarbon group):

Esters are formed through a process called dehydration because one water molecule is produced. An alcohol and an organic acid react together to form an ester and water. The monomers lose atoms that form the water, that's the dehydration part. After the hydrogen and hydroxide ions are removed from the monomers, they bond together to form a polymer, that is the ester.

The chemical name of an ester starts with the alcohol. You use the prefix of the alcohol to indicate the number of carbons, and then add the suffix, -yl. The second part of the name comes from the organic acid used. Use the prefix to indicate the number of carbons present, drop the -oic acid and replace it with -ate.

Examples:

1. Hexanol and acetic acid yields hexyl acetate
2. Ethanol and Butanoic acid yields Ethyl Butanoate.

III. Community Connections

Many American Indian tribes have been using natural items for making fragrances. They use fragrances to scent baskets, bowls, and trays. They burn incense to smudge and purify places. Inuit tribes use violet scent for incense and for fragrance that they place among their clothes. Sage is used to clear sacred space and is used before most ceremonies. Pinon is used to raise spirits, vanilla is used for peace and relaxation, lemongrass fragrance is used in ceremonies, and Fir needle extract is used for protection and healing.

IV. **Safety Considerations**
1. Always wear safety eye protection.
2. Some of the organic acids and alcohols may irritate skin and eyes. Flush with water for at least 15 minutes if needed and inform instructor.

V. **Materials**

Equipment	Samples and Reagents
Graduated cylinder, 10 mL	Ethyl alcohol, methyl alcohol,
Beakers, 2 - 400 mL	n-butyl alcohol
Eye droppers	Acetic acid, Butyric acid,
Hot plate	Lactic acid, Salicylic acid
Test tubes, small and rack	Sulfuric acid
Test tube holder	Water, distilled
	Test tube holder

VI. **Procedures**
1. Set up a cold-water bath by filling a 400 mL beaker about half full with water and ice.
2. Set up a boiling water bath by filling a 400 mL beaker about half full of water and place on hot plate.
3. Gather 5 test tubes and label them 1 through 5.
4. Add 2 ml of one of the alcohols (record smell in table) and 2 ml of one of the organic acid (record smell in table) listed in the chart below into a test tube. *You may have different alcohols and organic acids than listed below, modify as needed.)*
5. Add two drops of sulfuric acid to the test tube.
6. Loosely stopper the test tubes and place in the boiling water for 10 minutes.
7. Remove the test tube from the boiling water bath with a test tube holder and place in the ice bath.
8. Open the stopper and using a waving motion (wafting) smell and identify the ester.
9. Record the smell of the original alcohol, as well as your new product, and the name in the table below.
10. Repeat steps 4-9 with four other alcohols and their partnered organic acids as provided by your instructor.
11. Clean and return your glassware, return or discard left over materials according to your instructor, and leave your lab area clean.

Data Table

Test tube	Alcohol	Organic Acid	Product	Odor
1	n-butyl alcohol Odor:	Acetic acid Odor:		
2	Ethyl alcohol Odor:	Lactic acid Odor:		
3	Methyl alcohol Odor:	Salicylic acid Odor:		
4	Ethyl alcohol Odor:	Butyric acid Odor:		
5	Other Odor:	Other Odor:		

VII. Conclusions

1. Include your data table with answers in your report.
2. What are some common esters you use, are aware of, and/or have stories from your community and/or elders?
3. Summarize your results based on the lab objectives, your data, and an authentic application of the data *(5-10 sentences)*.

PROPOSAL FOR END OF SECOND SEMESTER PROJECT

The last lab is a student project (#26). Students are to connect chemistry, a community connection (Appendix B), and their own personal interests. Possible ideas for projects might be a short story, a film, a photo journal, a play, a costume, beadwork, a dance, a game, an experiment, a research paper … there are no limits as long as you can connect your project with chemistry!

A short proposal of your selected topic and preferred format (i.e. visual media, research, experiment) is due at midterms. Your instructor will provide you with requirements for this proposal.

20. SYNTHESIS OF ASPIRIN AND WINTERGREEN OIL

I. Objectives

1. To synthesize aspirin and oil of wintergreen, both of which are esters
2. To test the purity of the products

II. Facts to Know

Aspirin Background

The earliest recorded cure for pain cures, headache cures, and fever cures symptoms is attributed to a Greek named Hippocrates in about 400 BCE. Hippocrates administered willow leaf tea to ease pain of childbirth. He noted that some people found relief in a powder made from willow bark and leaves. The Ebers papyrus, an Egyptian medical text, circa 3000-1500 BCE, referred to willow as an anti-inflammatory and pain reliever for non-specific aches and pains. Moving forward, numerous references of using willow bark and leaves for medicinal purposes abound.

Salicin   Salicylic Acid   Acetylsalicylic Acid
              (Spirin)   (Aspirin)

In 1828, German pharmacy professor Johann Buchner isolated the active ingredient from willow bark, a bitter-tasting yellow crystalline substance, and he called it salicin. The next year, the French chemist Henri Leroux greatly improved his procedure. Swiss and German researchers discovered that Salicin was also found in the meadowsweet flower, *Filipendula ulmaria*. In 1838, the Italian chemist Raffaele Piria added a strong base to some of his pure salicin and succeeded in purifying the products. One product was glucose and he oxidized the other product to create salicylic acid. Some doctors used his salicylic acid to cure arthritis and rheumatism but many patients found it to be too bitter to the taste.

The most famous complaint about salicylic acid's bitterness occurred in 1898 when Felix Hoffman's father said that he wouldn't take any more of those bitter pills to cure his rheumatism. Luckily for his father, Hoffman was a chemist at the small Bayer Company in Germany. Hoffman searched the literature for other pain-relieving compounds and discovered that, in 1853, French chemist Charles Frederic Gerhardt neutralized its bitter taste by acetylating it to create acetylsalicylic acid (today known as aspirin). Gerhardt believed in the purity of scientific research, however, and did not patent his discovery. Hoffman mixed up a batch of the acetylsalicylic acid for his father and his father was very happy with it. For over a year, Hoffman tried to convince the company to sell this product. In 1899, they did and aspirin has remained the best-selling drug in the world ever since.

Wintergreen Background

William Procter Jr. in the USA and Auguste Andre Thomas Cahours in France made methyl salicylate using purified salicin and proved it was the same as wintergreen oil in 1842 and 1844, respectively. Today, wintergreen oil is used as a pain-reliever, flavoring of food, and as an insecticide.

Methylsalicylate
(Wintergreen Oil)

The WebMD (2018) list of medicinal uses of wintergreen oil include, arthritic treatment, asthma, fever, flatulence, inflammation, menstrual cramps, nerve pain, ovarian pain, pleurisy, and stomach pains. It is also applied to the skin to help relieve muscle pain. It is used as a flavoring agent for food, candies, teas, and some pharmaceutical products. Wintergreen is used in organic pesticides for controlling ants and other crawling insects.

The Aspirin Synthesis Reaction that you will perform

Salicylic Acid
(Spirin)

Acetic Anhydride

$+CH_3COOH$

Acetylsalicylic Acid
(Aspirin)

The Wintergreen Oil Synthesis Reaction that you will perform

Salicylic Acid
(Spirin)

$+CH_3OH \longrightarrow$

Methanol

$+H_2O$

Methylsalicylate
(Wintergreen Oil)

III. Community Connections

The Mohawk and Ojibwa Indians were known to chew on wintergreen leaves for the flavor and to reduce aches and pains. American colonists adopted that practice and then began adding it to their candles for the scent. They extracted the oil by squeezing the leaves. Modern day extraction of the oil is done by distillation.

Salicin and salicylic acid are found in many species of the Salicaceae family, in particular, willows (*Salix spp*.) and poplars (*Populus spp.).* These trees provided Native American Indians with medicinal treatments. For example, Aspen (*Populus tremuloides*) bark or roots was macerated (meaning chewed until soft) and applied as a poultice or salve to wounds, cuts, and abrasions. Others used a decoction of aspen bark or roots as a cold remedy, stomach pain, heartburn, and gynecological treatments. Balsam poplar (*Populus balsamifera*) was used as a salve, poultice, or wash for wounds, bruises, an antirheumatic treatment, and general illnesses. White poplar (*Populus alba*) was used for colds. Bark from the red willow (*Salix bonplandiana*) was made into a tea for fevers. Weeping willow (*Salix babylonica*) bark infusions were used for intestinal tract issues, laryngitis, and as a general tonic as well as used as a wash or poultice for dermatological problems.

IV. Safety Considerations

1. Wear eye protection at all times.
2. Sulfuric acid and acetic anhydride are extremely harmful to skin and eyes. Rinse the area immediately.
3. Methanol and ethanol are flammable. Methanol is toxic when inhaled or ingested even in very small quantities.
4. Acetic anhydride is very corrosive and irritating. It must be used in the hood. Immediately wash your skin or eyes if they come in contact with acetic anhydride.
5. Be careful not to inhale salicylic acid. It will irritate your throat.

V. Materials

Equipment	Samples and Reagents*
Beaker, 600 mL	salicylic acid; 7 g
Graduated cylinder, 10 mL	acetic anhydride; 8 mL
Buchner funnel	90% ethanol; 20 mL
Ice	1% ferric (III) chloride; 5 mL
Erlenmeyer flask, 125 mL	methanol; 5 mL
Side Arm Flask, 250 mL	6 M sulfuric acid; 5 mL
Ring stand & clamps	95% Ethanol
Hot plate in the hood	*per group of students
Melting point apparatus	

VI. Procedure

A. Aspirin Synthesis and Analysis

1. Label a 600-mL beaker with your name.

2. Create a "water bath" by adding 300 mL water to your beaker and start warming it on an electric hot plate located in the hood. The desired temperature is 70 °C.

3. Accurately weigh out 6 g salicylic acid. After you've placed a sheet of weighing paper on the balance, press the "tare" button to negate the mass of the paper. Then, weigh out the salicylic acid to the nearest 0.01 g.

4. Accurately measure 8 mL of acetic anhydride in a 10-mL graduated cylinder. Pour this into a 125-mL Erlenmeyer flask.

5. Add the solid salicylic acid to the liquid acetic anhydride. Swirling the flask and then carefully add in 10 drops of concentrated sulfuric acid.

6. Make sure the water bath temperature is 70 ± 5 °C and then place the flask into it. Keep warming it with occasional swirling until the salicylic acid completely dissolves (it takes about 15 minutes).

7. Use tongs to remove the flask from the water bath (but keep your water bath running for use in Step 9).

8. Cool the flask contents by placing it into some ice cold water. Crystals of acetylsalicylic acid should form in your flask within 10 minutes. If they don't form within 5 minutes, you should gently scratch the inside of the flask with a glass stirring rod to "seed" the crystals. Wait a full 10 minutes to make sure that all of acetylsalicylic acid has crystallized.

9. Assemble your vacuum filtration apparatus. A Buchner funnel is made of porcelain and has holes on its flat surface to allow liquid to flow through it rapidly. (It was invented by Eduard Buchner who won the 1907 Nobel prize for discovering alcoholic fermentation in the absence of live yeast). Place a piece of filter paper on top of the holes and wet its surface with some water. Place the Buchner funnel with its rubber neckpiece into a Side Arm Flask and make sure that all connections are tight. Attach one end of some thick-walled rubber tubing to the Side Arm of the flask. Attach the other end to an aspirator (a side arm on a water tap). When the tap water is running, it creates a vacuum inside your Side Arm Flask, which will rapidly draw any liquid through the Buchner funnel.

10. Collect your acetylsalicylic acid crystals using your vacuum filtration apparatus. First, empty the contents of your 125-mL Erlenmeyer flask into the bowl of the Buchner funnel.

11. Then, "rinse" out your Erlenmeyer by adding some ice water into it and emptying this into the Buchner funnel.

12. Finally, remove any unreacted acetic anhydride from your crystals by "washing" them with some ice cold water. Allow the crystals to dry by keeping the aspirator on for 5 more minutes.

B. Aspirin Recrystallization

13. Add 20 mL of 95% ethanol to a 100 mL beaker and scrape your wet crystals into it.
14. Warm your beaker on your 70 °C water bath until the crystals dissolve.
15. At the same time, warm 50 mL of water in another beaker.
16. Pour your ethanolic solution into the warm water.
17. If crystals form immediately, you will need to warm the solution again.
18. Bring the beaker to your bench and cover it with a watch glass.
19. Allow it to cool for 5 minutes during which time the acetylsalicylic acid should begin recrystallizing.
20. Don't dismantle your water bath yet.
21. Place the beaker on ice for 10 minutes more to complete the recrystallization process.
22. While waiting, move on to Part D.
23. Dump the solution from the side arm flask of your vacuum filtration apparatus into the drain.
24. Collect the aspirin crystals using the vacuum filtration apparatus just as you did in Step 8.
25. Remove the filter paper with its crystals from the Buchner funnel.
26. Gently scrape all of your crystals onto a piece of dry paper and let them dry in the air for a while.
27. There is probably uncrystallized acetylsalicylic acid in the filtrate located in the side arm flask of your vacuum apparatus.
28. If there is more than 30 minutes remaining in the class, you should collect this filtrate in a small beaker and heat it on the hot plate in the hood until its volume is reduced to about one-third of its original volume (~20 mL).
29. Repeat Steps 11 to 14 to collect the new crystals on the same filter paper as the first crop.

C. Aspirin Analysis
Aspirin Yield:

30. First, weigh a clean dry watch glass to the nearest 0.01 g.
31. Next, transfer the crystals to the watch glass and weigh it to the nearest 0.01 g. The difference is the mass of the acetylsalicylic crystals to the nearest 0.01 g.

32. Calculate the percent yield for your reaction. Propose two reasons it is different from 100%.

$$\% \text{ yield} = \frac{actual\ yield}{theoretical\ yield} \cdot 100$$

33. Theoretical yield calculations require a balanced chemical equation and molecular weight.

$$C_7H_6O_3 + C_4H_6O_3 \rightarrow C_9H_8O_4 + C_2H_4O_2$$

Salicylic	Acetic		Aspirin	Acetic Acid
Acid	Anhydride			

Molecular weight: $C_7H_6O_3$: 136 g/Mol $C_9H_8O_4$: 180 g/Mol

7-C: 7 X 12 = 84   9-C: 9 X 12 = 108

6-H: 6 X 1 = 6   8-H: 8 X 1 =   8

3-O: 3 X 16 = 48   4-O: 4 X 16 = 64

Calculation of Theoretical yield

$$6\ g\ C_7H_6O_3 \cdot \frac{1\ Mol\ spirin}{138\ g\ spirin} \cdot \frac{1\ Mol\ aspirin}{1\ Mol\ spirin} \cdot \frac{180\ g\ Aspirin}{1\ Mol\ aspirin} = \underline{\hspace{3cm}}\ g\ Aspirin$$

The Ferric Chloride Test: After you are done weighing your crystals, you should test it for purity. This test will determine whether your compound has a phenol. It will react with any remaining salicylic acid.

34. Add a couple drops of 1% ferric (III) chloride to some crystals of your aspirin. If the crystals are yellow, that indicates the crystals are pure because the ferric ion will add a yellow color; if the color is light purple, your aspirin is slightly impure; if the color is dark purple, your crystals are more impure. The purple color is determined by the other functional groups attached to the phenol.

Substituted Phenol   **Ferric-Phenol Complex**

The melting point test: Every solid has a characteristic melting temperature and melts over a narrow range of temperatures.

35. Crush some of your crystals and push them into a pile without using your fingers.

36. Push the open end of a capillary tube into the pile of crushed crystals until you get a significant amount of them into the tube.

37. Invert the capillary tube and gently bump it to get the crystals to settle into the bottom of the tube.

38. Place the tube into the melting point apparatus and watch very closely for the moment at which the crystals first begin to melt and the moment at which all of the crystals have melted.

39. Record the two temperatures as your melting temperature range.

D. Wintergreen Oil Synthesis and Analysis

40. Place 0.2 g of salicylic acid in a medium-sized test tube. It is not necessary to do this very accurately because you will not be able to weigh the product oil to determine yield.

41. Add 5 mL of methanol to the test tube and swirl until it dissolves.

42. Carefully add 5 drops of 6 M sulfuric acid to the mixture.

43. Heat the test tube in your 70 °C water bath for 15 minutes.

44. Remove and allow the test tube to cool to room temperature. Now, you may dismantle your water bath.

45. Carefully use your hand and waft the test tube scent toward your nose. Do not inhale directly!

    The Ferric Chloride Test: This test will determine whether your compound has a phenol, which it should.

46. Add a couple drops of 1% ferric chloride to a portion of your oil.

    A positive test reveals a purple color. The color is determined by the other functional groups attached to the phenol.

47. Clean and return your glassware, return or discard left over materials according to your instructor, and leave your lab area clean.

VII. Conclusions

1. Report the data you were asked to record in the procedures and the results of your calculations.

2. What was the role of sulfuric acid in the reaction? Consider the reaction equation when you answer this question.

3. What was the purpose of adding the ethanol to the crystals?
4. Why did the crystals form faster in the cold solution than in the warm solution?
5. Why did you test for phenol groups in your wintergreen oil product? Draw the structure of a phenol.
6. Explain how salicylic acid, acetylsalicylic acid, and methyl salicylate can all be analgesics. Be sure to refer to the structures of these molecules when you answer this question.
7. Name several natural sources of salicylic acid. Add any that you are familiar with from your family or community history.
8. Summarize your results based on the lab objectives, your data, and an authentic application of the data *(5-10 sentences)*.

VIII.   References

1. Connelly, D. (2014). A history of aspirin. The Pharmaceutical Journal. Retrieved from https://www.pharmaceutical-journal.com/news-and-analysis/infographics/a-history-of-aspirin/20066661.article.
2. Densmore, F. (1974). *How Indians Use Wild Plants for Food, Medicine & Crafts. (formerly called Uses of Plants by the Chippewa Indians).* Dover Publications, Inc. New York.
3. Easyecoblog. (2016). Winter Green Pest Control Tips-Ants, Organic Pesticides. Retrieved from https://www.easyecoblog.com/455/winter-green-pest-control-organic-pesticides/.
4. Eschner, K. (2017). Aspirin's Four-Thousand-Year History. Smithsonian.com Retrieved February from https://www.smithsonianmag.com/smart-news/brief-history-aspirin-180964329/.
5. International Aspirin Foundation. (2018). Aspirin Timeline. International Aspirin Foundation. Retrieved from https://www.aspirin-foundation.com/history-of-aspirin/aspirin-timeline/.
6. Moerman, D.E. (1990). *Native American Ethnobotany.* Timber Press. Portland, Oregon.
7. WebMD. (2018). Wintergreen. Retrieved from https://www.webmd.com/vitamins-supplements/ingredientmono-783-wintergreen.aspx?activeingredientid=783&activeingredientname=wintergreen.

21. THE SCIENCE OF SOAP MAKING

I. Objectives

1. To make a sample of soap from animal fat or vegetable oil by alkaline hydrolysis
2. To test how well the soap works in different types of water

II. Facts to Know

Saponification is when fats or oils are treated with aqueous hydroxide to convert the fatty acid esters into long chain carboxylates, which is how soap making begins. Soap depends on the fats and oils which are used in the saponification as well as other material such as fragrances, preservatives, colorants, and exfoliants, which are added during soap making. Other soap methods use animal fat such as tallow (rendered beef fat) or lard (rendered pork fat), but most craft-oriented books favor the use of vegetable fats and oils for reasons of animal rights to cosmetic properties.

Soap has been made by trial and error methods since before recorded history. In fact, some historians of chemistry have suggested that soap making was the first chemical reaction to be used by humans. The hypothesis is that some of the fire ash (known to be a strong base today and rich in KOH) came in contact with the meat fats (known to be triacylglycerides today) and water to produce a foamy and soapy substance.

$$\text{Fats} + \text{Ash} + \text{water} \rightarrow \text{Foamy and soapy stuff}$$
$$(\text{Triacylglycerides} + \text{KOH} + \text{water} \rightarrow \text{Fatty acids} + \text{glycerol})$$

The earliest evidence for soap production are Babylonian urns that date to 2800 BCE and that contain a soap-like substance. The earliest recorded recipe for soap is found on Babylonian tablets dated 2200 BCE. They say that cassia oil and ashes were boiled in water to produce a material that was used to clean cotton and wool fibers that were woven into textiles. The reason that they wanted to "clean" those fibers was so that they could dye them. Greasy fibers do not retain dyes nearly as well as clean ones. Soap was used in this way for millennia. It wasn't used for personal hygiene until much later.

One of the first manufactures of soap was during the first century A.D. is when a modest soap factory was built in Pompeii. During the Middle Ages, cleanliness of the body or clothing was not considered important, but an interest in cleanliness emerged again during the eighteenth century, when disease-causing microorganisms were discovered.

In 1813, Michel Eugène Chevreul became the first chemist to analyze the chemical nature of fats. He used lye (NaOH) to break the ester bonds of fats into glycerin plus "fatty acids." He also named many of the fatty acids, such as butyric, caproic, oleic, and stearic.

The Ester Hydrolysis Reaction

| Ester | Hydroxide | Carboxylate | Alcohol |

The Saponification Reaction

Triacylglyceride (tripalmitoylglyceride)

$+ 3 OH^-$

glycerol   Fatty acids (palmitic acid in its palmitate form)

Both plants and animals store their energy in the form of fats called triacylglycerides. These molecules have three fatty acids forming esters with glycerol (another common name is glycerin; the IUPAC name is propane-1,2,3-triol). Fatty acids without any double bonds are called saturated fatty acids. They have a low melting point and are solids at room temperature. Lards and shortenings are triacylglycerides composed of saturated fats.

Fatty acids with one double bond are called monounsaturated fatty acids. Oleic acid is the most common one. It has 18-carbons and a cis double bond located at the 9th carbon from its carboxylate end. From the structure, you can see that the cis double bond makes a "kink" in the backbone of the fatty acid. Remember that double bonds are planar and cannot rotate whereas single bonds can rotate through a variety of conformations. The kink in the backbone prevents the oleic acids from packing together. Since they cannot pack together, the melting temperature is very low compared to fully unsaturated fatty acids. As a consequence, they are liquids at room temperature. Soybean oil is 24% oleic acid.

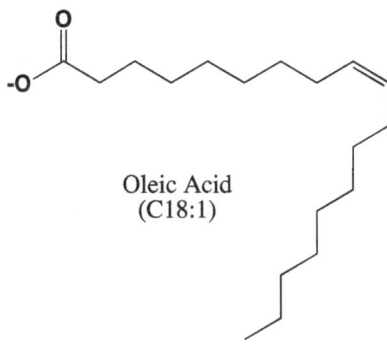

Oleic Acid
(C18:1)

Fatty acids with two or more double bonds are called poly-unsaturated fatty acids. Linoleic and Linolenic are the most common ones. Both have 18 carbons and cis double bond located at the 9th and 12th carbons from its carboxylate end. Linolenic also has one at the 15th carbon. Again, the double bonds prevent packing of the fatty acid side chains and these have very low melting temperatures. Soybean oil has 54% linoleic (C18:2) and 7% linolenic (C18:3) fatty acids in its triglycerides. Canola Oil has been engineered to have the highest percentage unsaturated fats; its composition includes 62% oleic (C18:1), 22% linoleic (C18:2), and 10% linolenic (C18:3).

Throughout history, soaps were made from animal fat, or lard. Today, most soaps are either made from vegetable oils or are synthetic detergents. These vegetable and synthetic soaps made their first appearance in the mid-1800s. In this experiment, you will saponify various oils and fats to establish the type of soap that is produced.

Saponification is when fats or oils are treated with aqueous hydroxide to convert the fatty acid esters into long chain carboxylates, which is how soap making begins. Soap depends on the fats and oils which are used in the saponification as well as other material such as fragrances, preservatives, colorants, and exfoliants, which are added during soap making. Other soap methods use animal fat such as tallow (rendered beef fat) or lard (rendered pork fat), but most craft-oriented books favor the use of vegetable fats and oils for reasons of animal rights to cosmetic properties.

III.   **Community Connections**

Native Americans used certain types of leaves, roots, berries, and barks that formed soapy lathers, which removed dirt from clothes. For example, soapwort and yucca plants were used as soaps by the West Coast and plains Indians. Other saponaceous plants found in North America include buffalo gourd *(Cucurbita foetidissima)*, yucca root *(Yucca glauca)*, and soap root *(Chlorogalum pomeridianum)* (2018). Several of the following plants were used by American Indians to make soap:

1. Buffalo gourd, _Cucurbita foetidissima_, was used for its medicinal properties more than its saponaceous properties. The seeds were used as a bleach for clothing. The Omaha made an analgesic from the roots pulverized in water.

2. Soap root, *Chlorogalum pomeridianum,* also known as amole, is native to the west coast, primarily California and Oregon. Also, soap root was used for food, and medicinal purposes. The brown fibrous bulb contains a toxin but when cooked is safe to eat.

3. Soapwort, *Saponaria officinalis,* was native to Europe but has been introduced worldwide and is most commonly known. Extract from the leafy stems and rhizomes contains the saponin for soap making. It is also known as "Bouncing Bet."

4. Soapweed Yucca, *Yucca glauca,* was a multiple-purpose plant for Native American Indians. The roots provided the saponin for soap-making, other parts of the plant were also important. The leaves were used for weaving in to ropes, baskets, mats, or sandals. The Zuni people boiled the seed pods for food. It is also a traditional medicinal plant for many Native American Indian tribes.

IV. Safety Considerations

1. Wear eye protection at all times.
2. Sodium Hydroxide (NaOH) is a corrosive toxin. It is harmful to skin, eyes, and clothing because it dissolves them. Rinse immediately after contact.
3. Ethanol is flammable. If you place ethanol near a hot flame, it will burn. Ethanol has a high vapor pressure, which means that invisible ethanol gas is present in the air above any liquid ethanol. This gas is particularly prone to burn when it comes near a flame.
4. Cyclohexane is flammable and should be kept away from open flames, sparks, and heat. It may irritate skin and eyes as well.

V. Materials

Equipment	Samples and Reagents
Beaker, 250 mL	Lard, margarine,
Buchner funnel	olive oil, coconut oil, mineral oil
Erlenmeyer flask, 125 mL	cyclohexane
Filter paper	95% ethanol
Graduated Cylinder, 50 mL	20% sodium hydroxide
Heating Plate	5% calcium chloride
Ice	Saturated sodium chloride
Stopper for 125 mL flask	Assorted commercial hand soaps
Stirring rod, glass	Universal indicator paper
Side Arm Flask, 250 mL	
Test tubes, 5, and test tube rack	

VI.   Procedures
Saponification (aka Soap Synthesis)

1. Add one of the following materials to a 150-mL beaker:
   10 mL vegetable oil, 5 g lard, or 10 mg margarine. In each case, weigh out the mass of your sample to within 0.01 g. If you use vegetable oil, weigh your (~10 mL) using a tared beaker.

2. Carefully measure 10 mL 95% ethanol in a 25 mL graduated cylinder and add to the sample. Melt the mixture on the hot plate in the hood. Stir with a glass rod.

3. Remove your beaker from the hot plate and bring it to your bench but don't let it cool too much. Carefully measure 12.5 mL of 20% NaOH solution and slowly add it to your ethanolic fat solution while constantly stirring with a glass rod.

4. Heat the solution using a hot plate while stirring constantly. Do not leave your reaction unattended. Heat until it becomes a pasty mass. After it has solidified, you should check that the alcohol has disappeared by wafting the vapors to your nose.

5. Bring your beaker to your bench and allow it to cool until it is warm to the touch, which may take as long as 15 minutes.

6. Add 50 mL of saturated sodium chloride while stirring. The fatty acids will form micelles and will float to the top of the sodium chloride solution. Allow the solution to cool to room temperature or place it on an ice bath to encourage all of the fatty acids to form micelles.

7. Assemble your vacuum filtration apparatus-a Buchner funnel. Place a piece of filter paper on top of the holes and wet its surface with some water. Place the Buchner funnel with its rubber neckpiece into a Side Arm Flask and make sure that all connections are tight. Attach one end of some thick-walled rubber tubing to the Side Arm of the flask. Attach the other end to an aspirator (a side arm on a water tap). When the tap water is running, it creates a vacuum inside your Side Arm Flask, which will rapidly draw any liquid through the Buchner funnel.

8. Turn on the water source. First, use your glass rod to prevent the soap from leaving your beaker and pour the liquid portion of your soap solution through the filter paper. Then, pour the soap portion onto the filter paper.

9. While the vacuum is still running, you may wish to "rinse" your beaker out using a small cup of ice water and then collect that as well. Allow the vacuum to run for five more minutes in an attempt to "dry" your soap.

10. Transfer the soap to a piece of paper towel to dry. Weigh the mass of your product to within 0.01 g.

11. If time, test your soap with commercial soaps available in the lab.

12. Clean and return your glassware, return or discard left over materials according to your instructor, and leave your lab area clean.

13. Calculate the % yield of soap assuming that the mass of the reagents and products doesn't change very much.

$$\% \text{ yield} = \frac{actual\ yield}{theoretical\ yield} \cdot 100$$

$$\% \text{ yield} = \frac{mass\ of\ soap}{mass\ of\ lard, margarine, or\ oil} \cdot 100$$

% yield= 100% × (mass of soap)/(mass of lard, margarine, or oil).

Soap Analysis

14. Dissolve about one-fourth of your soap in 40 mL water in a 125-mL Erlenmeyer flask, stopper it, and shake to form an emulsion that you will use in the following tests.

15. The Solubility Test. Separately label three test tubes as "soap", "water," and "cyclohexane." To each labeled tube, add 10 drops soap emulsion, 10 drops water, or 10 drops cyclohexane. Next, add three drops of mineral oil to each of the three test tubes, stopper them, and shake for 10 sec. Record your observations.

16. The Alkalinity Test. Use a piece of universal indicator paper to test the pH of your emulsion. If the paper turns red, the solution is acidic (pH < 7). If the paper turns blue, the solution is basic (pH > 7). Record your observations.

17. The Hard Water Test. Add 3 mL soap emulsion to a test tube labeled "$CaCl_2$." Add five drops of 5% Calcium Chloride to the test tube. Record your observations.

18. Clean and return your glassware, return or discard left over materials according to your instructor, and leave your lab area clean.

Data Table

Soap Sample	Solubility Test			Alkalinity Test	Hard H_2O Test
	Soap	Water	Cyclohexane	pH	$CaCl_2$
Lab-produced					

Making soap at home

Home samples: Lard + KOH, lard + leached wood ash or wood-ash paste, an oil with lye, Crisco with lye, borax for whitening, or food coloring if making soap at home. You could also try this with one of the plants discussed in the background section. Use the references within the table with pictures of the plants.

Procedures

1. Melt grease (animal grease), about ½ lb. of lard on a hot plate.
2. Add 50 grams of lye to 120 mL water (wash hands, use gloves). Use a wooden spoon.
3. Pour lukewarm melted grease into lukewarm lye solution stirring gently 15 minutes until batter is thick like honey.
4. If meat drippings were used for the fat, strain the melted drippings through cloth (for best texture soap). You can boil grease in twice as much water, then skim off the pure grease (optional).
5. Soap will have a light tan color and clean, fresh odor. To lighten color, add borax while stirring. If desired, add vegetable coloring or perfume scent (earth oil) when mixture has batter consistency.
6. Pour into cloth-lined box (cardboard box).
7. After 24 hours, it can be cut and stored.

VII. Conclusions

1. Show your results from the yield calculations. What is your hypothesis to explain the yield? What could you do to bring the yield closer to 100%?
2. Why is ethanol added to the lard, margarine, or vegetable oil?
3. Why is NaCl added to the reaction?
4. Did you test your soap with commercial soaps? Explain how you tested the soaps and your results.
5. Have you made your own soap at home or have seen it made? If yes, briefly explain what you used and results.
6. Summarize your results based on the lab objectives, your data, and an authentic application of the data (5-10 sentences).

VIII. References

1. Guana, F. J. (n.d.). *Soap Plant (Chlorogalum pomeridianum (DC.) Kunth)*. Retrieved from https://www.fs.fed.us/wildflowers/plant-of-the-week/chlorogalum_pomeridianum.shtml.

2. Haddock, M. (2007). *Buffalo Gourd*. Kansas Wildflowers & Grasses. Retrieved from http://www.kswildflower.org/flower_details.php?flowerID=35.

3. Hilty, J. (2017). Soapwort. Weedy Wildflowers of Illinois. Retrieved from http://www.illinoiswildflowers.info/weeds/weed_index.htm#soapwort.

4. Homestead Arts (2018). *Soap Lore*. Retrieved from http://homesteadarts.org/soaplore.html.

5. McDonald, C. (n.d.). *Soapweed Yucca (Yucca glauca)*. Retrieved from https://www.fs.fed.us/wildflowers/plant-of-the-week/yucca_glauca.shtm.

6. Outdoor self-reliance. (2013 July 23). *Buffalo Gourd*. Retrieved from http://outdoorselfreliance.com/buffalo-gourd/.

7. USDA Forest Service (n.d.) *Soaps*. Retrieved from https://www.fs.fed.us/wildflowers/ethnobotany/soaps.shtml.

22.TRANSESTERFICATION OF OILS IN NATIVE PLANTS

I. **Objectives**
1. To produce a biodiesel from a nut or seed.
2. To determine the percent oil found in the nut or seed.

II. **Facts to Know**

The transesterification process involves replacing one ester with a different ester. Vegetable oils are triglycerides which is a glycerin molecule connected via ester bonds to three fatty acid molecules (see figure below). The triglyceride portion of the oil is removed and attached to the alkyl group of an alcohol (the carbon and hydrogen part) to form fatty acid alkyl esters, or in other words, biodiesel.

Biodiesel is a cleaner burning renewable fuel than diesel. Biodiesel is mixed with diesel and can be burned in unmodified diesel engines at any concentrations.

Every type of vegetable oil and animal fat has different composition of fatty acids which determine its physical-chemical properties. The oils in nuts and sunflower seeds are predominantly triglycerides, which are trans-esterified to the methyl esters of the fatty acids. For more information about triglycerides see the soap making lab background information. A catalyst increases the rate of a chemical reaction. In this process, a base catalyst is used. This transesterification process produces a crude biodiesel as is it contaminated with methanol, base, glycerin, and possibly soap.

The generic reaction is:

oil	alcohol		biodiesel	glycerol

$$H_2C-OCOR_1$$
$$HC-OCOR_2 \quad + \quad 3\ CH_3OH \quad \xrightleftharpoons{catalyst} \quad R_1COO-CH_3$$
$$H_2C-OCOR_3$$

$$R_1COO-CH_3$$
$$R_2COO-CH_3 \quad + \quad H_2C-\underset{OH}{\overset{H}{C}}-CH_2$$
$$R_3COO-CH_3 \qquad\qquad OH \quad OH \quad OH$$

R1, R2, & R3 are hydrophobic end of fatty acids

III. **Community Connections**

Sunflowers are indigenous to western North America. American Indians grew sunflowers as far back as 3000 B.C.E. in what is now identified as Arizona and New Mexico, although evidence suggests sunflowers were domesticated in eastern North America 3500 B.C.E.

Sunflower seed was used for food in a variety of ways-ground or pounded, or as seeds (National Sunflower Association, n.d.). There are even references of squeezing the seeds for oil in the making of breads. Some referred to sunflower as the fourth sister in reference to the three sisters of squash, beans, and corn.

Sunflower meal was used to enhance the health of women, particularly pregnant or nursing women. The babies of nursing women who consumed sunflower meal were healthier than those of mothers who did not. Sunflowers were used a remedy for pulmonary troubles and lethargy. It also was used to treat snakebites. The oil was used on skin and hair as well (Trombley, n.d.).

Non-food uses included using the stalk as a building material. Purple pigment extracts were used for dyeing textiles, body paints, and decorations.

IV. Safety Considerations

1. Always wear goggles.
2. No open flames: n-hexane is very flammable.

V. Materials

Equipment

Graduated cylinder, 50 or 100 mL
Coffee grinder or mortar & pestle
Erlenmeyer flasks, 3 250 mL
Beakers, 2 - 250 mL
Stir plate with stir bars, if have one
Buchner or Funnel & filter paper
Side arm flask, rubber tubing
Ring stands with support rings, clamps
Condenser tubes
Separatory funnel
Hot plate for distillation
Distillation flasks

Samples and Reagents (per group)

Sunflower seeds in hull, 160-200 g, preferably a high-oil yielding variety*
200 mL hexane
1.2% KOH/CH_3OH mixture (50 mL)
Distilled water

*High oil content nuts or seeds or commercial vegtables oils may be used instead

VI. Procedures

We are thankful for the bounty of food that we are able to use in this lab.

1. Mass between 100 grams of raw sunflower seeds, hulls included to within 1.0 g; record the mass in the data table.
2. Grind the massed seeds. If you end up with a peanut-butter paste, you have ground too much. If using a mechanical grinder, select medium grind.
3. Add the ground sunflower seeds to a flask, containing 150 mL of n-hexane. Stir the mixture for 30 minutes.
4. Add 50 mL 1.2% KOH/CH_3OH solution. Allow this to stir for another 30 minutes.

148

5. Filter to remove the hulls and seeds; wash hulls and seeds with an additional 2-10 mL n-hexane as needed.
6. Place the mixture into a Separatory funnel. Allow this mixture to separate into two phases, then drain the bottom phase off into a beaker. Set this phase aside. You will discard it at the end of the lab.
7. Wash the remaining phase in the Separatory filter with 10 mL of hot H_2O, shake several times to intermix the liquids. Let sit 10 minutes before draining off the bottom phase in to a beaker. Set this phase aside for now.
8. Set up distillation apparatus: hot plate, flask, condensation tube, beaker for collection, using ring stand supports to stabilize, as demonstrated by instructor.
9. Look up the boiling point for n-hexane and record.
10. Distill the remaining phase; you will have two products, what is left in the distillation flask and the condensate collected in the beaker.
11. Determine the boiling point of your products: the condensate and what is left in the distillation flask.
12. Clean and return your glassware, return or discard left over materials according to your instructor, and leave your lab area clean.

Data Table:

Mass of sample:	g
Boiling point of n-hexane from reference	° C
Boiling point of recovered n-hexane	° C
Boiling point of products (methyl esters)	° C

VII. Conclusions

1. Include your data table in your lab report.
2. Why is n-hexane used as an extraction solvent and not water?
3. Why did we need the reference n-hexane boiling point information?
4. Why does the mixture separate into two phases after esterification of the oil?
5. Why is the top layer from the Separatory funnel washed with water?
6. Calculate percent yield of your esterification product. Show results here. Does it matter how much beginning sample you use?
7. To what general class of organic compounds is the oil classified?
8. What general class of organic compounds is the products of the reaction?

9. Summarize your results based on the lab objectives, your data, and an authentic application of the data *(5-10 sentences)*.

VIII. **References**:

1. Cole, R. S. (n.d.). *What are the Prospects for Energy Futures on Tribal Lands?* Retrieved from http://nativecases.evergreen.edu/collection/cases/what-are-the-prospects.

2. Lambert, L. (n.d.) *Alberta's Oil Sand and the Rights of First National Peoples to Environmental Health.* Retrieved from http://nativecases.evergreen.edu/collection/cases/albertas-oil-sands-and-the-rights-of-first-nations-peoples-to-environmental-health.

3. National Bio-diesel Board. (2018). *Welcome.* Retrieved from http://nbb.org/.

4. National Sunflower Association. (n.d.) *History.* Retrieved from https://www.sunflowernsa.com/all-about/history/.

5. Native Languages of the Americas website. (2015). *Native American Sunflower Mythology.* Retrieved from http://www.native-languages.org/legends-sunflower.htm.

6. Saul, K. M. (n.d.) *Should the Confederated Tribes of Warm Springs Invest in a Woody Biomass Co-Generation Facility?* Retrieved from http://nativecases.evergreen.edu/collection/cases/should-biomass-invest.

7. Trombley, J. (n.d.). *Foods Indigenous to the Western Hemisphere.* Retrieved from http://www.aihd.ku.edu/foods/sunflowers.html.

8. Turtle Mountain Community College. (2006). RISE Experiments.

9. Utah Biodiesel Supply. (2017). *Biodiesel tutorial Videos.* Retrieved from https://www.utahbiodieselsupply.com/tutorialvideos.php#smallbatch4.

23. EXTRACTION OF CAFFEINE FROM YERBA MATE

I. Objectives

1. To extract caffeine from Yerba mate
2. To compare Yerba mate extract with a caffeine standard

II. Facts to Know

Caffeine and other Alkaloids

Caffeine is an alkaloid, that is an organic amine. The prefix, 'caffe,' comes from the coffee plant and the suffix, 'ine,' identifies it as an amine molecule. The chemical formula for caffeine is $C_8H_{10}N_4O_2$. Alkaloids are produced by many plants as a secondary metabolite, giving the plant a bitter taste to animals. This serves as an anti-predatory characteristic.

Caffeine

In addition to their bitter taste, many alkaloids have pharmacological effects. Some alkaloids such as strychnine are neurotoxins. Others such as nicotine and caffeine produce a mild, pleasant stimulation when taken in low doses. The large group of opium alkaloids includes codeine, heroin, and morphine, remain the best pain-killing agents known. The list of alkaloids is very long but one other interesting category is the hallucinogens that include lysergic acid, the precursor for d-lysergic acid diethylamide (LSD)

Caffeine is found naturally in tea, coffee beans, and kola nuts. It is also added to carbonated beverages, chocolate, even some nonsteroidal anti-inflammatory drugs (NSAID), and can be bought as an over-the-counter stimulant like No-Doz. Undesirable side effects of caffeine are increased heart rate, nervousness, restlessness, and insomnia. You can become addicted to caffeine. The amount of caffeine per serving in several common household products is shown below:

Item Consumed	Amount of Caffeine per serving (mg)
Coffee	80 – 125 / 1 cup
Coke	45 / 12 oz.
Excedrin	33 - 65 / tablet
Mountain Dew	45 / 12 oz.
No-Doz	100 /tablet
Tea	30 – 75 /1 cup

All of these alkaloids are isolated from their natural sources by extraction. That is why these compounds are often called "natural products." Today, you will learn both solid-liquid and liquid-liquid extraction procedures. You will determine the success of your extractions either by thin layer chromatography or by melting point determination.

Extraction Procedures

Liquid-liquid and solid-liquid extractions require solvents with low boiling points for easy removal by evaporation. Other important properties are: the solvent does not react with any of the substances present, the second liquid should be immiscible with water; it should be nontoxic; it should be low cost; and it should appreciably dissolve the target substance much more than does water. For instance, if a molecule is somewhat soluble in water, then you won't be able to completely extract it from the water solution.

Thin Layer Chromatography

Thin layer chromatography combines the best of the two types of chromatography that you've already learned. The silica from the column chromatography is sprayed in a thin layer onto some plastic or glass to form a sheet like paper chromatography. You use it just like paper chromatography, which means that it is an analytical method and is not used to isolate large amounts of sample.

One important difference between paper chromatography and thin layer chromatography (TLC) is that it is possible to embed useful material either in the plastic backing or beneath the silica layer. Today, you will use TLC sheets that contain an embedded fluorescent material. When you shine an ultraviolet (UV) light on the TLC sheet, it will emit a yellowish light. If the sample that you apply to the sheet absorbs UV light, then it will prevent the embedded material from absorbing the light and it will appear as a dark shadow in the middle of yellowish light. Caffeine is able to absorb the UV light.

III. Community Connection

The Aztec Indians in South America were the credited with the first use of a caffeinated beverage. Montezuma is said to have consumes as many as 50 cups a day of this hot drink made from cacao leaves (Turtle Mountain Community College, 2006).

Another caffeinated drink called "mate" was popular along the eastern coast of South America. It was brewed from *Ilet paraguariensis*, a type of holly. Further north, American Indians used the leaves and stems of the yaupon holly, *I. vomitoria*, for a hot drink called "black drink." The yaupon holly, also called the "Christmas berry" tree or "cassena" is native along the eastern coast and Gulf coast of North America (Moerman, 2010).

This black drink was used daily during the village councils' deliberations especially for important council meetings. It was believed to purify the drinker, removing their anger and falsehoods. It was prepared by special village officials. After the meetings were over, those who attended the meetings would purge themselves by vomiting (Hudson, 1979). After the southeastern tribes were removed to Oklahoma, the tribes had to turn to other sources as *I. vomitoria* does not grow that far west.

The west coast region of North America hosts their own Yerba varieties, Yerba Santa, Y. Mansa, and Yerba Buena. Several different species of plants are categorized as Yerba Santa: *Anemopsis calfornica, Eriodictyon caflifornicum. E. lanatum, E. trichocalyx*; Yerba Buena: *Clinopodium douglasii*, formerly *Satureja douglasii*; Yerba Mansa: *Anemopsis caflifornicum* (Moerman, 2010).

IV.   Safety Considerations

1. Always wear safety goggles.
2. Ammonium hydroxide, dichloromethane, and chloroform may irritate your lungs if inhaled. Keep them in the hood as much as possible. In the lab, keep them covered.
3. Acetone is flammable—keep it away from flames
4. Sodium carbonate is a strong base that can dissolve your skin
5. Sodium sulfate is a skin irritant.
6. Caffeine is slightly toxic in its pure form and is readily adsorbed through the skin.
7. Do not shine the UV light into your eyes while using the UV Lamp.

V.   Materials

Equipment	Samples and Reagents
Beaker, 50 mL & 250 mL	Yerba mate
Capillary tubes	Pure caffeine, 3 g
Centrifuge	Coffee extract, 100 mL
Erlenmeyer Flask, 50, 250 & 500 mL	Tea extract, 100 mL, optional
Filter paper	Sodium carbonate (Na2CO3), 2 g
Funnel	Dichloromethane (CH2Cl2), 25 mL
Hot air blower -optional	Anhydrous sodium sulfate (Na2SO4), 5 g
Hot Plate	Distilled water
Melting point apparatus	6 M NH4OH, 3 mL
Test tubes, 6 medium	Chloroform, 10 mL 2 TLC sheets with
TLC chambers	fluorescent indicator
UV lamp	TLC solvent of 1:1 acetone: chloroform

VI.   Procedures

We are thankful for the bounty of food that we are able to use the yerba mate, coffee, and tea in this lab.

Liquid-liquid caffeine extraction

A solid-liquid extraction was prepared for you. Yerba mate was brewed (20 grams of yerba mate in 250 mL distilled water) by filling a large beaker with 10 tea bags, adding 200 mL distilled hot water, waiting 10 minutes for the extraction, decanting the "tea" into a second larger beaker, compressing the tea bags, and then repeating the procedure with 100 mL hot water.

1. Fill a 500- mL Erlenmeyer flask with either *100 mL* cold yerba mate extract or cold tea extract. *(Identify which sample was used, the yerba mate extract or a regular tea extract)*

2. Add *2 g of sodium carbonate* to the flask and swirl the flask until the sodium carbonate dissolves. This raises the pH of solution to deprotonate the dark-colored tannic acids so that they will be more soluble in the water phase than in the dichloromethane phase.

3. Add *25 mL of dichloromethane* and vigorously swirl the flask for 10 minutes. **Do not shake the mixture**. You do not want to form an emulsion this time because you won't be able to centrifuge to separate the liquids. This is an important step in the reaction. It is essential that you swirl for the entire 10 minutes.

4. Allow the mixture to stand for about 5 minutes and it will separate into 2 layers. There should be dark top layer and a clear bottom layer. You don't want the top layer, but it is in the way and you need to remove the top layer carefully. *Dichloromethane at 1.33 g/mL is denser than water.*

5. First, remove most of the top layer into another beaker using a pipette.

6. The filter the remaining layer and liquids. Fold filter paper into a cone shape and place it into a glass funnel. Fit the funnel onto an Erlenmeyer flask. Wet the filter paper with distilled water.

7. Gently pour the remaining portion of your bottom liquid layer into the funnel. The excess water will drain through, but the dichloromethane solution of caffeine will remain on the filter paper.

8. Use a pipette to transfer the dichloromethane solution to a 50 mL Erlenmeyer flask.

9. Add a *small scoop of anhydrous sodium sulfate* to it. It will absorb the remaining water traces.

10. In the meantime, label a clean, dry 50-mL beaker with your initials and determine its mass to within 0.01 g.

11. Transfer the solution to the tared 50-mL beaker, leaving behind any solid material.

12. Evaporate the dichloromethane on a warm hot plate in the hood until a milliliter of liquid is left. ***Do not allow it to boil!*** Allow the beaker to cool in the hood for a couple of minutes. This will allow the last remainder of dichloromethane to evaporate and leave behind solid residue of caffeine.

13. Reweigh your beaker to calculate the mass of the caffeine.
14. If you have access to melting point equipment, use it to find the melting point of your sample (using the crystals from above) and the known caffeine sample.
15. Clean and return your glassware, return or discard left over materials according to your instructor, and leave your lab area clean.

Data Table

Sample	Mass, g
Beaker, dry, clean & empty	
Beaker +solid residue	
Crystals (*Subtract empty beaker mass from beaker + residue)*	
Original yerba mate	20*

*Starting mass from prepared yerba mate extract., which may vary based on preparation of the tea sample.

Percent Recovered

The percent of caffeine on the yerba mate package was 1%. * Let's calculate the amount of caffeine you recovered:

$$\% \text{ recovered} = \frac{\text{mass of crystals}}{\textit{mass of original yerba mate}} * 100$$

Solid-liquid caffeine extraction

1. Prepare the caffeine standard. Label two test tubes: #1. yerba mate (tea) and #2. caffeine.
2. Mass 0.1 g of the recovered caffeine crystal from Part A.
3. Mass 0.1 g caffeine.
4. Extract the caffeine. Add 3 mL of 6 M NH4OH and 1 mL chloroform to each tube. Use a stirring rod to mix the solutions thoroughly. In this extraction, it is OK to form an emulsion while you stir because you will centrifuge next to break them up. An emulsion is a suspension of liquid droplets in another liquid. The ammonium hydroxide raises the pH of solution to deprotonate the any acids in the pill so that they will be more soluble in the water phase than in the chloroform phase. The chloroform is a nonpolar solvent that does not dissolve in water. Polar and charged compounds dissolve in the water. Hydrophobic compounds dissolve in the chloroform. Because water and chloroform don't mix, you have just done a solvent-solvent extraction of caffeine from a pill.

155

5. Before you centrifuge your samples to fully separate the two liquids, you need to balance each sample with one of its exact mass that contains water. You do this by placing a beaker on a balance, pressing the "tare" button to set mass to zero, recording the mass of your sample tube plus liquid, remove that tube and replace it with an empty one, and filling that tube with enough water that its mass equals your first tube. Be sure to label your balance tubes to keep track of them

Sample	Mass, g
Tube 1 + liquid	
Balance Tube 1 + water	
Tube 2 + liquid	
Balance Tube 2 + water	

6. Centrifuge your tubes for 4 minutes at modest speed to separate the layers.
7. Collect the caffeine layer (which is the lower layer). Use a Pasteur pipette to transfer the lower layer to a separate test tube. Squeeze the pipette bulb before you place the pipette tip into the liquid. *Chloroform at 1.73 g/mL is denser than water at 1.00 g/mL so it forms the bottom layer.*
8. Obtain your first thin layer chromatography sheet (TLC). Use a pencil to draw a thin, horizontal line that is 1 cm from the bottom on each piece of chromatography paper. This is your chromatographic origin. *(If you use a pen to draw the line, the pen ink will dissolve in the solvent and chromatograph as it runs up the paper.)* Use the pencil to place two X's along your chromatography paper. Below each X, use the pencil to indicate which sample will be placed where.

Origin ⟶

Std Sample

9. Use a capillary tube (a very thin tube) to spot each of the samples onto your X's. Use a separate capillary tube for each sample. Dip the tube into the liquid, allow it to draw up into the tube, touch the tube to the paper and allow it to be drawn into the paper. This is called "spotting." Do not allow your spots to spread too far. If they do, they will overlap into the adjacent spot and confuse your ability to interpret the results. As you spot, be sure to remove your capillary tube when the spot reaches about 0.5 cm. When the spot dries, you can add some more solution to the X. You may use a hot air blower (like a hair blower) to dry the sample quickly between additions. *Be careful that you don't heat it too much though as unlike paper chromatography, the TLC sheets can melt because they have a plastic backing!*

10. Place your TLC plate in a TLC chamber containing a layer of 1:1 chloroform: acetone solution that is no deeper than 0.5 cm. *(can improvise if you do not have one).*

11. While waiting for the solvent to travel up the silica, you can complete your yield calculations and clean up from Part A.

12. After the solvent has traveled up at least 8/10 of the TLC, remove the TLC sheet from the solvent and draw a line to show where the solvent traveled. Then, place the strips aside to dry for about 5 minutes. Shine the light from a UV lamp onto your sheet. Do not look into the UV lamp. A fluorescent dye is embedded in the silica so the entire sheet should glow except where you have UV-absorbing material, such as caffeine. Use a pencil to draw circle around the perimeter of each dark spot. If one of your samples generated two spots in the chromatogram, you should label the spots A, B, etc.

13. Analyze the TLC plates. Record the distance that the fastest edge of each spot traveled and the distance that the solvent traveled from the origin.

14. Calculate the Rf values:

$$Rf = \frac{Distance\ spot\ moved\ from\ the\ origin}{Distance\ solvent\ moved\ from\ the\ origin}$$

Sample	Distance, cm
Yerba Mate Spot	
Caffeine standard Spot	
Solvent Front from Origin	

VII. Conclusions

1. Report your data, a data table would be sufficient.

2. How much caffeine did you extract from the coffee or tea? What was your percent yield?

3. What were the Rf values for the sample and standard in Part B. Use them to explain whether you were able to extract any caffeine from the pills. Explain how to determine the purity of your caffeine extract.

4. Tea also contains a molecule called theobromine. It is very similar to caffeine in its structure and its physiological effects. What is the structure of theobromine? Circle the part that is different from caffeine.

5. Pure caffeine is a white solid. Why might your product not be white?

6. What are two benefits and two detriments of caffeine consumption?

7. What are some beverages you know of from native plants? Do they contain caffeine?

8. Summarize your results based on the lab objectives, your data, and an authentic application of the data *(5-10 sentences)*.

VIII. References

1. Guana, F. J. (n.d.). *Yerba Buena (Clinopodium douglasii (Benth.) Kuntze).* Retrieved from https://www.fs.fed.us/wildflowers/plant-of-the-week/clinopodium_douglasii.shtml.

2. Guana, F. J. (n.d.). *Yerba Buena (Eriodictyon sp. Benth.).* Retrieved from https://www.fs.fed.us/wildflowers/plant-of-the-week/eriodictyon_sp.shtml.

3. Hudson, C. M. (1979*). Black Drink*. University of Georgia Press.

4. Moerman, D.E. (2010). Native American Food Plants. An Ethnobotanical Dictionary. Timber Press: Portland.

5. Natural Resources Conservation Service (n.d.) *Anemopsis californica (Nutt.) Hook. & Arn. yerba mansa.* Retrieved from https://plants.usda.gov/core/profile?symbol=anca10.

6. Natural Resources Conservation Service. (n.d.). *Illex vomitoria Aiton yaupon*. Retrieved from https://plants.usda.gov/core/profile?symbol=ilvo.

7. Coladonato, M. (1992). *Ilex vomitoria.* In: Fire Effects U.S. Department of Agriculture, Forest Service, Rocky Mountain Research Station, Fire Sciences Laboratory (Producer). Retrieved from https://www.fs.fed.us/database/feis/plants/shrub/ilevom/all.html.

8. Heck, C. & de Mejia, E. G. (2007). Yerba Mate Tea (*Illex paraguariensis*): a comprehensive review on chemistry, health implications, and technological considerations. *J. Food Sci. 72*(9), R138-151. DOI:10.1111/j.1750-3841.2007.00535.x

9. Turtle Mountain Community College. (2006). RISE Experiments.

24. THE EFFECT OF HEAT ON ENZYMES FOUND IN NATIVE AMERICAN FRUITS

I. **Objectives**

    1. To investigate the effects of heat on enzymes found in several fruits.

II. **Facts to Know**

    Enzymes are proteins that help bring about chemical changes. An enzyme's function is related to the 3-dimensional structure of its molecule.   Some fruits contain enzymes that break down the protein structure in meat. Meat tenderizers contain an enzyme found in fruit called papain, which comes from the papaya fruit. These enzymes can break down the protein found in gelatin, and interfere with its gelling process.

    Enzymes usually have optimal conditions in which they will be the most effective. Certain environmental factors such as temperature and pH, can interfere with the enzymes actions, causing it to be inactive.

III. **Community Connections**

    American Indians have used fruits and nuts as sources of food. During the harvest season, summer and fall, varieties of apples, chokecherries, varieties of plums, varieties of sandcherries, hazelnuts, and other edible fruits and nuts were gathered. These foods were eaten both fresh and dried. While edible by themselves, many of the dried foods were used for flavoring. Fruits and other plant parts such as leaves and roots were often used as medicine as well. They may have been used as a poultice or rub, or consumed as a tea.

    For example, rosehips, also known as rose haws, have been used widely by humans globally. Rosehips have been used for beverages, preserves, relishes, sauces, and snacks. The vitamin C content decreases after freezing. However, after a first frost, not a hard freeze, the rosehips taste sweeter due to the release of sugars from damaged cells or cell components. If the rosehips are collected in the mid-to-late fall, dried, and then stored, the Vitamin C content is higher than those collected after a hard freeze. High heats also decrease the vitamin C content whereas temperatures below boiling do not.

IV. **Safety Considerations**

    1. Always wear safety goggles
    2. Do not consume the food used in this lab during lab.

V. Materials

Equipment	Samples and Reagents
Graduated cylinder, 10 mL & 100 mL	One envelope of Knox gelatin
Containers, 24: test tubes, small beakers,	Approximately 10 grams of the
Culture dishes, or reclosable snack bags	following fruits: pineapple,
Hot plate or other source of hot water	blueberries, cranberries, kiwi, apples,
Beaker, 600 mL	chokecherries, rosehips
3 reclosable bags	Water, distilled

VI. Procedures

We are thankful for the bounty of food that we are able to use the fruits and vegetables in this lab.

1. Separate your fruit into three groups: fresh, frozen, and heated. Place each group into a zip lock bag.
2. Place one zip lock bag into the freezer, one in the beaker of hot water, and leave one bag out while you prepare the gelatin or for 15-30 minutes.

   Temperature of Freezer: _____ °C

   Room Temperature: _____ °C

   Temperature of hot water: _____ °C
3. Prepare the gelatin using less water than the standard recipe, approximately 200 mL of water.
4. Divide and number your containers into three sets of 1-8, F for frozen, H for hot, and R for room temperature and place the following into each.
5. Add 5 mL of gelatin to each container.
6. Stir the fruit and gelatin together.
7. Place the containers in the refrigerator.
8. Observe and record every 15 minutes. The experiment is done when containers numbered 1 set up, gelled completely. The tables are set up for 2 hours but you may adjust time.

Container number	Contents
1	Nothing
2	Blueberries
3	Chokecherries
4	Cranberries
5	Kiwi
6	Pineapple
7	Rosehips
8	Apples

9. Clean and return your glassware, return or discard left over materials according to your instructor, and leave your lab area clean.

Data Tables

Fresh Fruit *(fruit at room temperature)*

	15 minutes	30 minutes	45 minutes	60 minutes	75 minutes	90 minutes	105 minutes	120 minutes
1								
2								
3								
4								
5								
6								
7								
8								

Heated fruit *(temperature heated at: _____ °C)*

	15 minutes	30 minutes	45 minutes	60 minutes	75 minutes	90 minutes	105 minutes	120 minutes
1								
2								
3								
4								
5								
6								
7								
8								

Frozen fruit *(temperature frozen at: _____ °C)*

	15 minutes	30 minutes	45 minutes	60 minutes	75 minutes	90 minutes	105 minutes	120 minutes
1								
2								
3								
4								
5								
6								
7								
8								

VII.   **Conclusions**

1. Summarize your data in table format without repeating data tables from procedures.
2. Which fresh fruits had enzymes present that interfered with the protein in the gelatin?
3. Which heated fruits had enzymes present that interfered with the protein in the gelatin?
4. Which frozen fruits had enzymes present that interfered with the protein in the gelatin?
5. Based on your results, does heat affect the activity of enzymes in fruit? Explain your answer.
6. Based on your results, does cold interfere with the activity of enzymes in fruit? Explain your answer.
7. What is (are) the active ingredients in meat tenderizer?
8. What do you think would happen if you placed a mixture of meat tenderizer and gelatin each in hot water, ice water, and room temperature water, and then placed all three of them in a refrigerator? Support your answer with evidence based on this lab or experimentation at your home.

9. Summarize your results based on the lab objectives, your data, and an authentic application of the data *(5-10 sentences)*.

VIII.   Reference

1. Leahu, A., Damian, C., Oroian, M., Ropciuc, S., & Rotaru, R. (2014). Influence of Processing on Vitamin C Content of Rosehip Fruits. *Scientific Papers: Animal Science and Biotechnologies. 47*(1): 116 – 120.

2. Moerman, D.E. (2010). Native American Food Plants. An Ethnobotanical Dictionary. Timber Press: Portland.

3. Roman, I., Stanila, A., & Stanila, S. (2013). Bioactive compounds and antioxidant activity of *Rosa canina* L. biotypes from spontaneous flora of Transylvania. *Chem. Cent. J. 7*(73). doi: 10.1186/1752-153X-7-73.

4. Santos, P. H. S. & Silva, M. A. (2008). Retention of Vitamin C in Drying Process of Fruits and Vegetables-A Review. *Drying Technology. 26*(12). 1421-1437. Retrieved from http://ucanr.edu/datastoreFiles/608-216.pdf.

5. Turtle Mountain Community College. (2006). *RISE Experiments.*

25. DNA EXTRACTION

I. Objectives

1. To extract the DNA from a common food.
2. To estimate the concentration of DNA.
3. To determine the total yield and purity of the DNA sample.

II. Facts to Know

James Watson (b. 1928), Francis Crick (1916-2004), Maurice Wilkins (1916-2004), and Rosalind Franklin (1920-1958) determined the structure of deoxyribonucleic acid (DNA) in 1953. The first three scientists were awarded the Nobel Prize in physiology or medicine in 1962. Franklin had died of cancer in 1958, and the Nobel Prize is never awarded posthumously.

DNA is the basis for heredity, contains the patterns for constructing structural and functional (enzymes) proteins in the body. DNA consists of two chains twisted around one another to form a double helix, of alternating phosphate and sugar groups that are held together by hydrogen bonds between pairs of organic bases-adenine (A) with thymine (T), and guanine (G) with cytosine (C).

Nitrogen Base Components of Nucleic Acids

DNA only	RNA only
Thymine	Uracil

DNA & RNA		
Adenine	Guanine	Cytosine

Sugars and Phosphate Components of Nucleic Acids

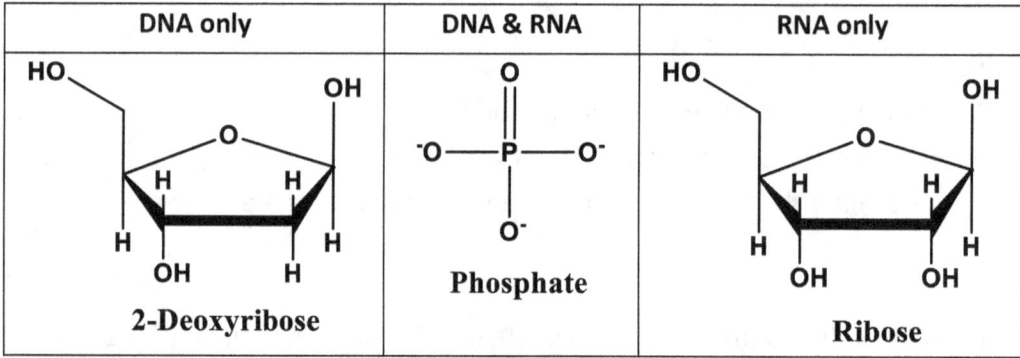

DNA only	DNA & RNA	RNA only
2-Deoxyribose	Phosphate	Ribose

The evidence for the helical DNA structure was based on Franklin's X-ray diffraction images of DNA fibers. She was the only trained chemist of the group, with a Ph. D. in physical chemistry. Wilkins and Crick were physicists and Watson was a zoologist.

For an individual of any species, the sequence of base pairs contains the coded messages that determine the potential characteristics of that individual. DNA is the template that stores information for recall as needed. It contains the code of life which is passed from one generation to another.

There are numerous commercial companies available that will identify your heritage from your DNA as well as check for marker genes for certain diseases such as diabetes and breast cancer. Some of these companies are associated with genealogy companies and others with health organizations. Other uses for genetic/DNA testing is parental testing and genetic "fingerprinting."

After extracting the DNA, the concentration may be estimated by measuring how much ultraviolet light it absorbs. DNA absorbs in the 190 to 320 nm region. Total yield may be calculated by taking into consideration the sample volume and its concentration. One problem with using ultraviolet absorption is that protein also absorbs in the same range. This is a problem because natural DNA is extracted with some bound protein. Purity of the DNA sample may also be determined by the estimated concentration and yield data. The A_{260}:A_{280} is a common measure of the purity of a DNA sample as absorbance at 280nm gives an estimate of protein contamination of the DNA extract. A measurement of the DNA sample at 320 nm provides a measure of background turbidity which also assists in determining the purity of the DNA sample.

III.   **Community Connection**

For genetic information directly relating to American Indians, see American Indian and Alaska Native Genetics Resource Center (http://genetics.ncai.org/tribal-enrollment-and-genetic-testing.cfm).

IV.   **Safety Considerations**

1. Always wear safety goggles.
2. Food used in this lab should not be consumed.

V.   **Materials**

Equipment	Samples and Reagents
Graduated cylinders 10 mL and 100 mL	Medium - Large Onions
Test-tubes, 5, and rack	Meat tenderizer
Knife	Colorless dish soap
Blender	70% Isopropyl alcohol, chilled
Beakers, 50, & 250 mL	Water, distilled
Ice bath for alcohol	
Funnel and filter paper	
Glass stirring rod	
Spectrophotometer & quartz cuvettes	
Centrifuge and tubes	

VI.   **Procedures**

We are thankful for the bounty of food that we are able to use the fruits and vegetables in this lab.

Extract the DNA

1. Cut a 2.5 cm square out of the center of the onion. Chop and place in the blender.
2. Dissolve 2.5 g salt in about 60 mL hot water.
3. Add salt solution to onions and blend on high 30 seconds – 1 minute. It is okay if there are still some onion chunks in the mixture.
4. Pour the mixture into a 250 mL beaker and add about 60 mL colorless dish soap.
5. Stir gently for 5 minutes. Attempt to form as few bubbles as possible. This step removes the lipids and membranes.
6. Filter the mixture.
7. Add 2.5 g meat tenderizer to filtrate and stir gently to mix. This step removes the proteins and enzymes.
8. Place 5 mL of filtrate into a test tube. This 5 mL is your total sample volume which you will need for your calculation.
9. Pour 5 mL ice cold isopropanol carefully down the side of the tube to form a layer. The layer of isopropanol should be about twice as thick as the onion filtrate.
10. Let the mixture sit for 2-3 minutes undisturbed until bubbling stops.
11. The DNA will float in the alcohol layer. Swirl a glass stirring rod at the interface of the two layers to see the small threads of DNA.

12. Spool the DNA onto a glass rod by gently swirling under and lifting up into the DNA mass to remove it from the test tube.
13. Allow the excess isopropanol to drip off or press it out gently against the side of the test tube. Place the spooled DNA into s a 50-mL beaker.
14. Wash the DNA with 15-20 mL of isopropanol, decant excess alcohol, place the DNA into a tube and centrifuge for 5 minutes. Remember you will need to prepare and place a blank tube of the same mass opposite your DNA sample test tube.
15. Repeat this washing and centrifuge step the times by re-suspending the pellet and rinsing.
16. Dissolve the pellet in distilled water for the UV spectrophotometry analysis.

Concentration and Yield

17. Clean the quartz cuvette with lint-free tissue. Rinse the cuvette chamber with a small amount of distilled water. Fill the cuvette with distilled water and place in the spectrophotometer. Record a blank measurement with this cuvette.
18. Measure the A_{220}, A_{260}, A_{280}, and A_{320} nm for each sample. If your spectrum is off-scale, then your sample is too concentrated and needs to be diluted further. Record your final dilution factor.
19. Calculate your concentration, total yield, and purity of your DNA sample:
    a. Concentration: A_{260} of 1.0 = 50 µg/mL of pure double-stranded DNA; if you did not dilute, omit that in your calculation.

    $$\text{Concentration}_{DNA} = (A_{260} \text{ reading} - A_{320} \text{ reading}) *(\text{dilution factor}) * 50 \text{ µg/mL}$$

    b. Total yield:

    $$\text{Yield}_{DNA} = \text{Concentration}_{DNA} * \text{total sample volume, mL}$$

    c. Purity:

    $$\text{Purity}_{DNA(260/A280)} = \frac{(A_{260} reading - A_{320} reading)}{(A_{280} reading - A_{320} reading)}$$

20. Clean and return your glassware, return or discard left over materials according to your instructor, and leave your lab area clean.

VII.   Conclusions

1. Report your results in an appropriate format.

2. Does your tribe use a blood quantum test to determine affiliation? If yes, describe how the blood quantum is determined. If no, choose a tribe that has such as requirement and describe it.

3. Summarize your results based on the lab objectives, your data, and an authentic application of the data *(5-10 sentences)*.

VIII.   References

1. Bardill, J. (n.d.) *Tribal Enrollment and Genetic Testing.* American Indian & Native Alaska Native Genetics Resource Center. National Congress American Indian. Retrieved from http://genetics.ncai.org/tribal-enrollment-and-genetic-testing.cfm.

2. Chem. 247.53 Biophysical Chemistry. (2010). *Isolation, spectrophotometric Analysis and Melting of Onion DNA.* Retrieved from https://chem247.files.wordpress.com/2007/09/chem-247-dna-lab.pdf.

3. Promega corporation. (2018). *How do I determine the concentration, yield and purity of a DNA sample?* Retrieved from https://www.promega.com/resources/pubhub/enotes/how-do-i-determine-the-concentration-yield-and-purity-of-a-dna-sample/.

4. Reusch, W. (2013 May 5) *Nucleic Acids.* Retrieved from https://www2.chemistry.msu.edu/faculty/reusch/virttxtjml/nucacids.htm.

5. U.S. National Library of Medicine. (n.d.) *The Rosalind Franklin Papers.* Profiles in Science. Retrieved from http://profiles.nlm.nih.gov/ps/retrieve/Narrative/KR/p-nid/183.

6. Science History Institute. (2018). *James Watson, Francis Crick, Maurice Wilkins, and Rosalind Franklin.* Retrieved from http://www.chemheritage.org/discover/online-resources/chemistry-in-history/themes/biomolecules/dna/watson-crick-wilkins-franklin.aspx.

26. SECOND SEMESTER PROJECT AND PRESENTATION

VI.   Objectives

1.   To research a community topic of student interest.
2.   To prepare a presentation, poster and oral, of results of research.

VII.   Facts to Know

Your learning is personal to you, and the content we talk about this semester will be meaningful to you in unpredictable ways. As we proceed through the semester, keep an eye out for topics and experiences that jump out to you, are related to the community topics listed in the Appendix B, and are related to chemistry. Think about how you might put your own spin on them for a creative project at the end of the course. You will have two lab periods to work on these with a third session for presentation your projects to the rest of the class.

Possible ideas for projects might be a short story, a film, a photo journal, a play, a costume, beadwork, a dance, a game, an experiment, a research paper … there are no limits as long as you can connect your project with chemistry! The creative project will be graded not only on the accuracy of the scientific information included, but also for originality and craftsmanship. A short proposal (roughly one paragraph) describing your plans for your project was due the week of midterms. Your lab instructor's approval was required for your project selection.

Your proposal should include the topic, the connection to chemistry, your chosen format for creating, and a list of the materials you think you need. If you do an experiment, you will have help with your lab design and collection of data. If you choose to do a research project, you will have help with research skills, paper format, and writing as well. You could do a research project and present in poster format instead of a research paper.

A rubric for your project with grading criteria will be provided once you have submitted your proposal and it has been approved by the lab instructor.

VIII.   Community Connections

After selecting a topic of interest from the community issues (Appendix B), you will research that topic as related to chemistry concepts. You may choose to conduct an investigation/experiment. Your instructor will help you with experimental design, equipment and materials, and collecting data. You will have the remaining weeks of lab time to work on this project. Presentations will be during finals week, during the last lab session.

IX.   Safety Considerations

1.   Depends upon your topic and research methods.

X. **Materials**

Research topic from community issues related to chemistry

Poster-minimum size, 24" x 36"

Internet

Equipment and Materials if conducting an experiment as needed

Procedure for Preparing a Poster:

1. **How to create a poster for a science meeting.** There are three major things to consider when you present your research findings. The first is that you have full understanding of your hypotheses, methods, research findings, and interpretation. The second is the visual aspects of the poster itself. The third is the oral presentation you give to people who stop by your poster during the conference to learn about your research. The following description provides guidelines and tips for the second aspect, which is preparing a visually exciting poster that effectively communicates your findings. There are also tips about the oral presentation.

   Poster Size. Before you begin, you should consult the meeting guidelines about the poster size. You can use a wide range of programs for creating your poster but each of them requires you to set the image size before you begin adding content. You have been provided with a choice of two PowerPoint documents as a starting point. The "6 Poster Layout" is set to 40 x 46 and contains a mostly blank field. The "PosterPresentations.com" is set to 36 x 48 so you may have to adjust the size but it contains many fields just waiting for you to plug stuff into them. You can also search the internet to find many other poster templates. In fact, there is an insightful, relatively amusing, but overly long blog about creating posters that you might want to check out: http://colinpurrington.com/tips/academic/posterdesign.

   Overall, the poster should attract attention and convey information, use no jargon, have no spelling errors, use large text, be well-organized, and simple. The main sections of a poster are: title, abstract, introduction, results, conclusion, references, and acknowledgments. Here are suggestions for each of these sections.

   a. **Title.** The title area takes up the top section of the poster. It is common to place your school logo in the upper left and upper right corners. The title has three lines. The first line of the title should be in a very large font size (such as 100 points) that stretches across the top. The title should be short and avoid abbreviations. The second line gives the authors names in slightly smaller font size (such as 80 points). The presenter's name is listed first. The third line of the title gives the authors' institutional affiliations in an even smaller font (such as 60 points).

b. **Abstract**. Usually located in the upper left corner below the title field. Every meeting lists the number of words you can use in your abstract, so you need to check that first. In fact, it is common to submit your abstract to the meeting, which is then used to determine whether you will be invited to give your poster. You are usually limited to 300 words or less to convince someone that your work is valuable. Therefore, the abstract has to state your purpose, goals, hypothesis, major finding, and possibly describe the major implications of your findings. Since the abstract is so short, you might consider writing it as follows:

    i.   *Opening Sentence*: Describe an important issue to grab the reader's attention. This can be one of our Community Topics.

    ii.   *Second Sentence*: Describe the long-range goal of your research area. A long-range goal addresses the issue you mentioned in the first sentence. For instance, describe a specific topic related to a Community Topic.

    iii.   *Third Sentence*: Justify your long-range goal. Describe why your specific project is likely to yield results. State why you chose to study your specific topic.

    iv.   *Next: State your hypothesis* (one sentence). State what instrumentation you used to carry out your research. Finally, describe the major findings.

    v.   Strong closing sentence: Explain how your results address your hypothesis and bring you closer to reaching your long-term goal.

c. **Introduction**. Usually located below the abstract, in which case the abstract and introduction take up the left-most fourth of your surface. The introduction describes the background that the listener would need to understand why you carried out your research. This would be a description of one of our Case Studies. It often includes a note about methods and instrumentation if they are special.

d. **Results**. Usually located in the middle section of the poster. This is the largest and most variable part of your poster. Use bold images with color highlights and font size 24 or larger. Make your graphs and tables first. As you add each new item to the poster, consider how you will use the images when describing your results at the meeting. The order of the images should help you describe the flow of ideas.

e. **Conclusions**. Usually located in the upper right corner below the title field. This is typically a short series of bullet points.

f. **References**. This is sometimes located at the end of the introduction unless you cite references in the Conclusions.

g. **Acknowledgments**. Usually in the lowest right-hand corner. List the funding source for the research and help of anyone who is not listed as a co-author.

2. **Tips for the oral presentation at a science meeting**. Every symposium chooses how long the oral presentations should be. For the purposes of this lab exercise we will choose 5 minutes. You will have only 1 min to describe the reason you did the research and 4 min to describe your research findings. Since everyone is best able to describe their research findings, this means you should focus on creating your 1-minute introduction and a very short take-home conclusion. Practice giving the oral several times out loud to yourself and several times to your friends. One last tip is to keep your finger or hand in place for several seconds when you are pointing to one of your images. It communicates a sense of purpose and calm to your listener as opposed to a jab in the general direction of the image which communicates you are in too great of a hurry.

Example of a Creative Project Grading Rubric

Grading Category	Points	Criteria
Connection	25%	25%: Elaborates significantly on chemistry & community topic 20%: Elaborates on chemistry & community topic 15%: Chemistry or connection lacking but not both 10%: Superficial chemistry connection 5%: Minimal chemistry connection 0%: No chemistry connection
Content	25%	25%: All chemistry information is accurate & current 20%: Most of the information is accurate & current 15%: Inaccuracies OR omissions 10%: Inaccuracies AND omissions 5%: Majority of information is inaccurate and incomplete 0%: Chemistry information is all inaccurate or missing totally
Creativity	25%	25%: Unique 20%: Best of similar projects 15%: Well done example of a category of products 10%: Average example of a category of projects 5%: Poor example 0%: Uncreative
Craftsmanship	25%	25%: Very well done, professional project 20%: Neat, well-done, effort is evident 15%: Nice looking but needs some corrections and refining 10%: Significant improvements to appear professional 5%: Little time & effort is evident 0%: No effort evident.

Appendices

Appendix A: REFERENCE TABLES

Table 1: The First 92 Atomic Elements[a]

Element No.	Symbol	Name	Atomic Mass	Element No.	Symbol	Name	Atomic Mass
1	H	hydrogen	1.008	37	Rb	rubidium	85.47
2	He	helium	4.0033	38	Sr	strontium	87.62
	Li	lithium	6.941	39	Y	yttrium	88.91
4	Be	beryllium	9.012	40	Zr	zirconium	91.22
5	B	boron	10.81	41	Nb	niobium	92.91
6	C	carbon	12.01	42	Mo	molybdenum	95.96
7	N	nitrogen	14.01	43	Tc	technetium	[98.00]
8	O	oxygen	16.00	44	Ru	ruthenium	101.1
9	F	fluorine	19.00	45	Rh	rhodium	102.9
10	Ne	neon	20.18	46	Pd	palladium	106.4
11	Na	sodium	22.99	47	Ag	silver	107.9
12	Mg	magnesium	24.31	48	Cd	cadmium	112.4
13	Al	aluminum	26.98	49	In	indium	114.8
14	Si	silicon	28.09	50	Sn	tin	118.7
15	P	phosphorus	30.97	51	Sb	antimony	121.8
16	S	sulfur	32.07	52	Te	tellurium	127.6
17	Cl	chlorine	35.45	53	I	Iodine	126.9
18	Ar	argon	39.95	54	Xe	xenon	131.3
19	K	potassium	39.10	55	Cs	cesium	132.9
20	Ca	calcium	40.08	56	Ba	barium	137.3
21	Sc	scandium	44.96	57	La	lanthanum	138.9
22	Ti	titanium	47.87	58	Ce	cerium	140.1
23	V	vanadium	50.94	59	Pr	praseodymium	140.9
24	Cr	chromium	52.00	60	Nd	neodymium	144.2
25	Mn	manganese	54.94	61	Pm	promethium	[144.9]
26	Fe	iron	55.85	62	Sm	samarium	150.4
27	Co	cobalt	58.93	63	Eu	europium	152.0
28	Ni	nickel	58.69	64	Gd	gadolinium	157.3
29	Cu	copper	63.55	65	Tb	terbium	158.9
30	Zn	zinc	65.38	66	Dy	dysprosium	162.5
31	Ga	gallium	69.72				
32	Ge	germanium	72.63				
33	As	arsenic	74.92				
34	Se	selenium	78.96				
35	Br	bromine	79.90				
36	Kr	krypton	83.80				

Element No.	Symbol	Name	Atomic Mass
67	Ho	holmium	164.9
68	Er	erbium	167.3
69	Tm	thulium	168.9
70	Yb	ytterbium	173.1
71	Lu	lutetium	175.0
72	Hf	hafnium	178.5
73	Ta	tantalum	181.0
74	W	tungsten	183.8
75	Re	rhenium	186.2
76	Os	osmium	190.2
77	Ir	iridium	192.2
78	Pt	platinum	195.1
79	Au	gold	197.0
80	Hg	mercury	200.6
81	Tl	thallium	204.4
82	Pb	lead	207.2
83	Bi	bismuth	209.0
84	Po	polonium	[209.0]
85	At	astatine	[210.0]
86	Rn	radon	[211.0]
87	Fr	francium	[212.0]
88	Ra	radium	[226.0]
89	Ac	actinium	[227.0]
90	Th	thorium	232.0
91	Pa	protactinium	231.0
92	U	uranium	238.0

[a]Note: Brackets indicate the element does not have natural isotopes (that is, all of its isotopes are unstable).

Table 2: All Metric Prefixes

Prefix	Pronunciation	Symbol	Sci. Not.	Long-hand and written notation
yotta	YAH tah	Y	10^{24}	1,000,000,000,000,000,000,000,000 (One septillion)
zetta	ZEH tah	Z	10^{21}	1,000,000,000,000,000,000,000 (One sextillion)
exa	EHK sah	E	10^{18}	1,000,000,000,000,000,000 (One quintillion)
peta	PEH tah	P	10^{15}	1,000,000,000,000,000 (One quadrillion)
tera	TEHR ah	T	10^{12}	1,000,000,000,000 (One trillion)
giga	GIHG eh	G	10^{9}	1,000,000,000 (One billion)
mega	MEHG eh	M	10^{6}	1,000,000 (One million)
kilo	KIHL uh	k	10^{3}	1,000 (One thousand)
hecto	HEK toh	h	10^{2}	100 (One hundred)
deka	DEK ah	D	10^{1}	10 (Ten)
deci	DEHS eh	d	10^{-1}	0.1 (One-tenth)
centi	SEHN tch	c	10^{-2}	0.01 (One hundredth)
milli	MIHL eh	m	10^{-3}	0.001 (One-thousandth)
micro	MY kroh	⍰	10^{-6}	0.000001 (One-millionth)
nano	NA noh	n	10^{-9}	0.000000001 (One-billionth)
pico	PY koh	p	10^{-12}	0.000000000001 (One-trillionth)
femto	FEHM toh	f	10^{-15}	0.000000000000001 (One-quadrillionth)
atto	AT toh	a	10^{-18}	0.000000000000000001 (One-quintillionth)
zepto	ZEHP toh	z	10^{-21}	0.000000000000000000001 (One-sextillionth)
yocto	YAHK toh	y	10^{-24}	0.000000000000000000000001 (One-septillionth)

Table 3: The *International System of Units*: Seven Basic Units

Property	SI unit	SI symbol
amount of substance	mole	mol
electric current	ampere	A
length	meter	m
luminosity	candela	cd
mass	kilogram	kg
temperature	Kelvin	K
time	second	s

Note: SI symbols do not have periods after them and there are no plural and singular versions. For instance, one meter and two meters are written 1 m and 2 m.

Table 4: Common Metric Prefixes Used in Chemistry

Prefix	Pronunciation	Symbol	Scientific Notation	Written Out
kilo	KIHL uh	k	10^3	1,000 (One thousand)
milli	MIHL eh	m	10^{-3}	0.001 (One-thousandth)
micro	MY kroh	μ	10^{-6}	0.000001 (One-millionth)
nano	NA noh	n	10^{-9}	0.000000001 (One-billionth)
pico	PY koh	p	10^{-12}	0.000000000001 (One-trillionth)

See Table 2 for a list of all metric prefixes

Table 5: Common Conversion Factors

Property	Foot-pound system	SI system
Energy	1 calorie (cal)	4.1868 joules (J)
Energy	1 Calorie (Cal) = 1000 cal	4186.8 joules (J)
Energy	1 kilowatt-hour (kW-h)	3.6 megajoules (MJ)
Length	1 mile	1.609344 kilometers (km)
Mass (or weight)	1 pound (lb) = 16 oz.	453.59237 grams (g)
Mass (or weight)	1 stone = 14 pounds (lb)	6.3502932 kilograms (g)
Mass (or weight)	2204.6226 pounds (lb)	1 metric ton (tn)
Pressure	101,325 atmospheres (atm)	1 pascal (Pa)
Pressure	6894.7573 pounds/square inch (psi)	1 pascal (Pa)
Volume	1 quart (qt)	0.94635295 pascal (Pa)
Volume	1 gallon = 4 quarts = 8 cups	3.7854118 liters (L)

Note: There is a fundamental difference between mass and weight. Mass is the quantity describing the physical matter within a substance. Weight is mass times the force of gravity. For example, the mass of an astronaut remains the same whether on Earth or the Moon but the astronaut weighs much less on the Moon because the Moon mass (and, therefore, gravity) is less than Earth's.

Table 6: The Alkanes

Name	#C	Composition	Molar mass[a]	Number of compounds with this generic name[b]
Alkane	n	C_nH_{2n+2}		
Methane	1	CH_4	16.0	1
Ethane	2	C_2H_6	30.1	1
Propane	3	C_3H_8	44.1	1
Butane	4	C_4H_{10}	58.1	2
Pentane	5	C_5H_{12}	72.1	3
Hexane	6	C_6H_{14}	86.2	5
Heptane	7	C_7H_{16}	100.2	9
Octane	8	C_8H_{18}	114.2	18
Nonane	9	C_9H_{20}	128.3	35
Decane	10	$C_{10}H_{22}$	142.3	75
Undecane	11	$C_{11}H_{24}$	156.3	159
Dodecane	12	$C_{12}H_{26}$	170.3	355
Tridecane	13	$C_{13}H_{28}$	184.4	802
Tetradecane	14	$C_{14}H_{30}$	198.4	1858
Pentadecane	15	$C_{15}H_{32}$	212.4	4347
Hexadecane	16	$C_{16}H_{34}$	226.4	10,359

[a]Molar mass (in u or g/mol) = 12.01 x n + 1.008 x (2n+2), where n = number of carbon atoms.
[b]The number of compounds does not account for stereoisomers, which we will encounter in a later chapter.

Table 7: Straight-chain alkane properties

Name	Boiling Point (°C)	Melting Point (°C)
Methane	-162	-183
Ethane	-89	-182
Propane	-42	-188
n-Butane	0	-138
n-Pentane	36	-130
n-Hexane	69	-95
n-Heptane	98	-91
n-Octane	126	-57
n-Nonane	151	-54
n-Decane	174	-30

Molar mass (in amu or g/mol). The *n* in the names means *normal* and indicates the carbons are connected in a straight chain.

Table 8: Energy for breaking bonds between pairs of atoms

Bond Name	Bond Symbol	Average Bond Energy	
Single Bonds			
Carbon-carbon single bond	C—C	82.4 kcal	345 kJ
Carbon-hydrogen single bond	C—H	99.1	415
Carbon-oxygen single bond	C—O	83.6	350
Hydrogen-hydrogen single bond	H—H	103	432
Hydrogen-oxygen single bond	H—O	111	464
Double Bonds			
Carbon-oxygen double bond, in most molecules	C=O	170	712
Carbon-oxygen double bond, in CO_2	C=O	187	783
Oxygen-oxygen double bond	O=O	119	498

Note: These energies are given per mole of bonds, which would be 6.02×10^{23} bonds. These values are reported in both kilocalories and kilojoules because you are undoubtedly familiar with food calories whereas scientists use joules. One kilocalorie (kcal) is the same energy as one food Calorie (note the capital C); 1 Cal = 1 kcal and 0.001 Cal = 1 calorie. There are two simple conversion equations that relate these two units: 1 calorie = 4.18 joule; and 0.239 calories = 1 joule.

Community Topics (Holistic/Interconnectedness)

Air Quality

Animal Habitat

Biopiracy

Climate Change (Trends, Historical Knowledge, Ecosystems)

Community Health (Genetics, GMOs, Food Sources)

Disease (Historical, New, Current)

Economic Development (Trust Lands, Environmental Racism)

Food Sovereignty

Medicinal Plants

Natural Resources (Soil, Land, Forests, Fire Management)

Oral Histories

Ownership/Stewardship

Renewable Energy (Solar, Wind, Compressed Wood Pellets)

Waste (Hazardous, Solids, Land Fills)

Water Sources (Natural Disasters, Remediation Programs, Metals, Testing, Policy, Watersheds)

Appendix **C. LIFE FLOWS: CONNECTING CHEMISTRY TO WATER**
by Bev DeVore-Wedding
Framing the Chemistry Curriculum
University of Nebraska and Nebraska Indian Community College

"I, as a Ho-Chunk child, was taught to respect that which is the giver and taker of life; Creator, water, fire."

LaVonne Snake, 2015

Introduction

Water is intricately woven into the creation stories of many different cultures, including American Indians. The following stories come from the American Indian Tribes associated with Nebraska Indian Community College, Little Priest Tribal College, and the Great Plains.

"When Earthmaker created the world he looked down and shed his tears into the void. These tears became the waters, which are the essence of the Waterspirit's being." (Dieterle, n.d.)

"According to one Dakota creation story, a creation story that 'figures prominently in Lakota/Dakota creation stories,' the sacred lake Mde Wakan is where the Dakota emerged as human beings into this world. The sacred lake is where the Lakota/Dakota people's primary (or first) Garden of Eden site is located, and it is the Garden of Eden from which they were forced out, and to which they will return. Mille Lacs Lake is where their (first) genesis site is located." (Dahlheimer, 2016).

"The Mdewakanton, 'those who were born of the waters,' have lived on Prairie Island for countless generations. Located in Southeastern Minnesota along the wooded shores of the Mississippi and Vermillion Rivers, Prairie Island is a spiritual place for our people." (Prairie Island Indian Community, 2018).

"There are seven groups of Dakota [Mdewakanton, Wahpekute, Wahpeton, Sisseton, Yankton, Yanktonai, and Teton]. There are seven stars in the constellation of Orion. We are the spirit beings from the constellation of Orion and those seven stars. The whole area is important to us because this is where we first came as spirit beings—to the confluence of the Mississippi and Minnesota rivers. We spread out from there becoming human beings as we spread out from there." (Dahlheimer, 2016).

"'The water from Coldwater Spring comes out from underneath the land and some of the spirit beings that arrived went into the water and they appeared on earth here and so became Dakotahs.' The original and authentic Dakota origin creation story actually says the Dakota's

*place of origin is near **Mde Wakan**, or "about the lakes at the head of the Rum River." Another Dakota origin creation says that after a flood some of the people entered into Mille Lacs Lake and lived underwater, and later emerged from the sacred Lake. Today's Dakotahs are descendants of the Mille Lacs Lake underwater people who emerged from the sacred lake as human beings into this world." (Dahlheimer, 2016).*

The Wakanda, or Water God   Yankton

"A man and his wife had only one child, they say, whom they loved very much. He used to go playing every day, they say; and one day he fell into the water. His father and mother and all his relations wailed regularly. His father was very sad, they say. He would not sleep within the lodge; he lay out of doors, without any pillow at all. When he lay on the ground with his cheek on the palm of his hand, he heard his child crying. He heard him crying down under the ground, they say. Having assembled all his relations, he spoke of digging into the ground. The relations collected horses to be given as pay; they collected goods and horses. Then came two old men who said they were sacred. They spoke of seeking for the child. An old man went to tell the father. He brought the two sacred men to the lodge. The father filled a pipe with tobacco. He gave it to the sacred men, and said, "If you bring my child back, I will give all this to you."

So they painted themselves; one made his body very black, the other made his body very yellow. Both went into the deep water. So they arrived there, they say. They talked to the wakanda. The child was not dead; he was sitting up, alive.

The men said, "The father demands his child. We have him; we will go homeward," they said. "You have him; but if you take him homeward with you, he shall die. Had you taken him before he ate anything, he might have lived. Be gone ye, and tell those words to his father."

The two men went. They arrived at the lodge, they say. 'We have seen your child; the wakanda's wife has him. We saw him alive, but he has eaten of the food of the wakandas. Therefore, the wakanda says that if we bring the child back with us out of the water, he shall die.'

Still, the father wished to see him. 'If the wakanda's wife gives you back your child; she desires a very white dog as pay.'

'I promise to give her the white dog'" said the father.

Again the two men painted themselves; the one made himself very black, the other made himself very yellow. Again they went beneath the water. They arrived at the place again.

'The father said we were to take the child back at any cost; he spoke of seeing his child.'

So the wakanda gave the child back to them; homeward they went with him. When they reached the surface of the water with him, the child died. They gave him back to his father. Then all the people wailed when they saw the child, their relation.

They plunged the white-haired dog into the water. When they had buried the child they gave pay to the two men.

After a while, the parents lost another child, a girl, in the same way, they say. But she did not eat any of the wakanda's food, therefore they took her home alive. But it was another wakanda who took her, and he promised to give her back if they would give him four white-haired dogs." (Judson, 2007).

Topic

Without water, life as we know it would not exist. Water flows through our communities in a variety of ways: human and livestock consumption, gardens and large-scale farming, native plants and animals, groundwater and rivers as sources for water use, treatment of wastewater, and the economics of water supply and demand.

Water makes up approximately 75% of the Earth's surface, about 70% of your body, and is essential to life (Atteberry, 2010). Humans can exist days, even weeks without solid food, but can only last up to 10 days without water depending upon the environmental conditions (Scientific American, 2002; Binns, 2012).

Water is essential to life. Our bodies are about 65-70% water. Water is the essential for the biochemical reactions which power and keep us alive. Dehydration, a reduction of water in our bodies, affects our functioning. Thirst is your body's warning you need water. By the time you feel thirsty, you have already reduced your body's water by about 2% (Maldarelli, 2017). Continued dehydration affects your cognitive and organ functioning (Maldarelli, 2017).

We use water daily, both directly and indirectly. Direct water use includes the consumption of water as a beverage or food, preparing food, and cleaning ourselves, our clothes, dishes, and our domicile. Indirect water use includes agriculture that is growing the food you consume, industry involved in processing and packaging that food, and manufacturing other items you use such as your computer, your vehicle, your clothing, and your toiletries. Globally, individuals require at least 20 to 50 liters (5- 13 gallons) per day but in the United States of America, individuals use 50 – 80 gallons daily (National Academy of Sciences, 2007; Environmental Protection Agency, 2015).

Polluted water isn't just dirty but can be deadly due to chemical pollutants themselves and also due to the sanitation quality. Water borne diseases cause the death of some 1.8 million people and tens of millions become ill by water-related ailments globally annually, most of which are easily preventable with clean water (National Academy of Sciences, 2007). Thus, water is essential for hydration, food production, and sanitation.

The United Nations considers access to clean water a basic human right. Even here in the United States of America, there are communities without clean water, or even without any water, such as the Navajo in northern Arizona and New Mexico (Bureau of Reclamation, 2013; Paskus, 2015; Smith, 2015). Every city in the USA has had water issues. For example, Macy NE had boil water alerts (Montag, 2013). Santee NE and Niobrara NE have had issues with flooding over the last 160 years (Niobrara, n.d.). The photograph (Figure C.1) shows the mineral deposit in a S. Sioux City drinking fountain. Is *that water* safe to drink? If you knew some chemistry, you would know those mineral deposits are calcium carbonate (white), manganese carbonate (green), and iron carbonate (red). That is the deposits indicate the water is rich in minerals and nothing more.

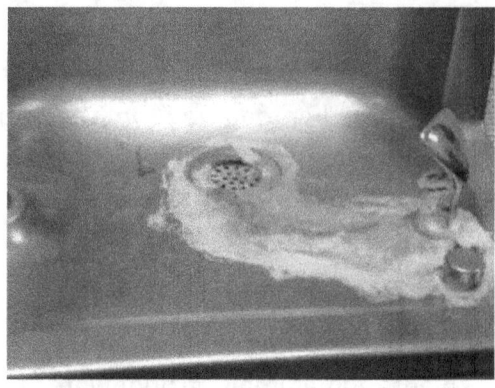

Figure C.1. Mineral deposits are calcium carbonate, indicating the drinking water is rich in minerals.

Problem

Is your local water safe to drink and use for agricultural purposes? What is the evidence to support the safety of your water?

Discussion Questions
Agriculture
1. How does water quality affect crop production? (Small and large scale farming)
2. Are native plants better adapted to the water regime and climate than the large-scale mono-culture crops?
3. What are the uses of native plants?
4. How does the nutritional value of indigenous plants compare to non-indigenous and invasive plants?
5. How does water affect soil quality?
6. What is the quality of the soil used for farming? How do you test soil quality?

Drinking water
7. Who tests the water to determine not only what is in it but if it is safe to drink?
8. What does the water quality report for your campus location report for violations, concerns, and water content?
9. How does your water quality compare with that of at least one other community in the NICC system or Little Priest Tribal College community? What are the differences? What are the similarities?

10. What are some of the possible water-borne or water-related diseases that might appear in your drinking water? How would you treat water to prevent to prevent these diseases?

Groundwater

11. What is the source of the groundwater?
12. How is groundwater recharged?
13. How do contaminants enter the groundwater system?
14. What are possible agricultural contaminants to groundwater?
15. What are future concerns about groundwater sources?
16. What are some of the possible water-borne or water-related diseases that might appear in the groundwater? How would you treat water to prevent to prevent these diseases?

Surface water

17. Who tests the water quality of lakes, ponds, and streams in your area?
18. Why would you want to test these surface water systems?
19. What contaminants could be found in surface water and where would those contaminants come from?
20. What are some of the possible water-borne or water-related diseases that might appear in surface water? How would you treat water to prevent to prevent these diseases?

Water Treatment

21. What happens to the water after is leaves your dwelling? What is the treatment process, if any?
22. Are there chemicals in the drinking water that is harmful to humans? Crops? Animals?
23. What stories have been passed on to you involving water?
24. What else would you like to know about the quality of your water?

Resources

Not sure where to look for water quality resources? Try your local town or county government's website. If they provide water, then they must test it and provide an annual water quality report. For example, S. Sioux City's (2018) town government website has a page dedicated to water and water quality, including the annual water quality report.

Nationally, there are several resources starting with the Environmental Protection Agency (FPA). They provide information not just on drinking water, but on wastewater treatment, monitoring and prevention of water pollution, bodies of surface water, and suggestions for individual participation in water quality monitoring.

Another organization, the Environmental Working Group (EWG) provides water quality databases in each state, pollutant reports, and consumer resources. For example, their databases provide information on all three towns with Nebraska Indian Community College's campuses, Macy, Santee, and S. Sioux City.

The Water Quality Association (WQA) (2018) promotes water quality globally. Its website provides basic water information, water contamination concerns and mitigation plans, and general resources.

The bibliography lists both works cited and additional resources for studying water, water chemistry, and human interactions with water. Enjoy your researching!

Bibliography

Atteberry, J. (2010, August 10). *"Why is water vital to life?"*. Retrieved from http://science.howstuffworks.com/environmental/earth/geophysics/water-vital-to-life.htm.

CBS News. (2015 August 13). EPA: High levels of toxic metal in Animas River water after mine spill. Retrieved from https://www.cbsnews.com/news/epa-high-toxic-metal-levels-in-animas-river-after-mine-spill/.

Dahlheimer, T. I. (2016). *A History of the Dakota People in Minnesota.* Retrieved from http://www.towahkon.org/Dakotahistory.html.

Dahlheimer, T. I. (n.d.). The Coldwater Spring Deception. Retrieved from http://www.towahkon.org/Coldwater.html.

Dieterle, R. L. (2015 June). *Water Spirits.* Retrieved from http://www.hotcakencyclopedia.com/ho.Waterspirits.html.

Enivornmental Protection Agency. (2015, April 9). *Water.* Retrieved from http://water.epa.gov/.

Enivornmental Protection Agency. (2015, August 21). *Standards and Risk Management.* Retrieved from http://water.epa.gov/drink/standardsriskmanagement.cfm.

Enivornmental Protection Agency. (2018). *Water Topics.* Retrieved from https://www.epa.gov/environmental-topics/water-topics.

Environmental Working Group. (2018). *EWG's Tap Water Databases.* Retrieved from https://www.ewg.org/tapwater/#.WuCmjojwaUl.

Farabee, M. (2001). *Structure of Water.* Retrieved from CHEMISTRY II: WATER AND ORGANIC MOLECULES: http://nld.df.uba.ar/biofisica/BioBook/BioBookCHEM2.html#Table of Contents.

Groundwater Foundation (1996, February). *The Basics.* Retrieved from http://www.groundwater.org/get-informed/basics/whatis.html

Letzler, R. (2017, November 29). *How long can a persion survive without water?* Retrieved from http://www.livescience.com/32320-how-long-can-a-person-survive-without-water.html.

Montana State Government. (2015). *Missouri Headwaters State Park*. Retrieved from
http://stateparks.mt.gov/missouri-headwaters/.

Judson, K. B. (ed.) (2007). Myths and Legends of the Great Plains. Retrieved from
http://www.gutenberg.org/files/22083/22083-h/22083-h.htm#Page_126.

Maldarelli, C. (2017 February 28). This is what happens to your body as you die of dehydration.
Retrieved from https://www.popsci.com/dehydration-death-thirst-water.

Maher, H. D. Jr. (2017). *Nebraska's Water Sources*. Retrieved from
http://maps.unomaha.edu/maher/GEOL1010/lecture9/lecture9.html.

Miller, J. (1999). *USGS Ground Water Atlas for the United States: Kansas, Missouri, and
Nebraska*. (U. S. Survey, Editor) Retrieved http://capp.water.usgs.gov/gwa/index.html.

Montag, M. (2013, February 23). *Omaha Tribe seeks federal help for Macy, Neb., water fixes*.
Retrieved from http://siouxcityjournal.com/news/local/a1/omaha-tribe-seeks-federal-
help-for-macy-neb-water-fixes/article_b0bc2346-9429-5464-9b2e-f00eca752c45.html.

National Academy of Sciences. (2007). *Why is safe water essential?* Retrieved from
https://www.koshland-science-museum.org/water/html/en/Overview/Why-is-Safe-
Water-Essential.html.

National Geographic. (2010 April). *Water Our Thirsty World. A special Issue.*

National Science Foundation. (2018 April 25). *A Special Report: The Chemistry of Water*.
Retrieved from https://www.nsf.gov/news/special_reports/water/.

Paskus, L. (2015). 'We're Going to Be Out of Water': Navajo Nation Dying of Thirst. Retrieved
from https://indiancountrymedianetwork.com/news/environment/were-going-to-be
-out-of-water-navajo-nation-dying-of-thirst/.

Prairie Island Indian Community (2018). Retrieved from http://prairieisland.org/.

Bureau of Reclamation (2013, July 16). *Upper Colorado Region*. Retrieved from
http://www.usbr.gov/uc/rm/navajo/nav-gallup/.

Schrage, S. (2015, August 17). *Study: Two major U.S. aquifers contaminated by natural uranium*.
Retrieved from
http://newsroom.unl.edu/releases/2015/08/17/Study%3A+Two+major+U.S.+aquifers+c
ontaminated+by+natural+uraniu.

Packer, R.K. (2018). *How Long Can the Average Person Survive Without Water?* Retrieved from
https://www.scientificamerican.com/article/how-long-can-the-average/.

Smith, N. L. (2015 August 19). Navajo Nation confiscates water tanks after mine spill. Retrieved
from https://www.daily-times.com/story/news/local/navajo/2015/08/19/navajo-
nation-confiscates-water-tanks-after-mine-spill/32380361/.

South Sioux City Nebraska. (2018). *Public Works Department.* Retrieved from
https://www.southsiouxcity.org/department/index.php?structureid=12.

United States Geological Society. (2018 January 29). The USGS Water Science School. Retrieved
from https://water.usgs.gov/edu/.

United States Geological Society. (2018). Water Resources. Retrieved from https://www.usgs.gov/science/mission-areas/water-resources.

United States Geological Society. (2018 April 16). *Water Quality*. Retrieved from
    http://water.usgs.gov/edu/waterquality.html.

Village of Niobrara (n.d.). *Niobrara*. Retrieved from http://www.niobrarane.com/history.html.

Water Quality Association. (2018). Water Basics. Retrieved from https://www.wqa.org/.

Water Policy International Ltd. (2002, December 9). *How long can the average person survive
    without water?* Retrieved from http://www.thewaterpage.com/live-without-water.htm.

Water Policy International Ltd. (2012). *How long can you live without water?* Retrieved from
    http://www.thewaterpage.com/live-without-water.htm.

Appendix D. STUDENT LAB REPORTS

Introduction

An essential component of doing science is communicating your data, analyses, and conclusions. Reports are the end result of doing science investigations and experiments. The first lab on safety, equipment and measurement report form sheets are provided as specific information is desired. The next ten labs (Density through Exothermic & Endothermic Reactions) lab report forms are provided for the student. The molar mass of butane lab begins the transition to students providing their own report form, with the remainder of the labs expecting the student to follow the report form template provided below.

Student Lab Report Expectations

The lab report form should include the student's name, class and section, partners if any, title of the lab as part of the heading. After that, the objectives of the labs should be stated as a reminder of what the lab was about and to assist the student in writing the lab summary. Students may repeat the safety considerations, materials, and procedures, but to save time and paper, if there were no changes to these in the lab, then simply stating that after these headings is sufficient. Data should be organized, preferably in a table or graph. If data recording tables are given, those can be copied and pasted into students' report forms. The conclusion is a summary of the lab. Questions have been provided on the student report forms or in the conclusions section of the lab that are to be answers. Numbering them as in the given report forms is preferably with a summary statement at the end. The summary itself will include one-three sentences addressing the data, what it implies, any unexpected results as well as expected results, without completely repeating the data. Another one-three sentences regarding any changes to the lab if any. Then three to five sentences regarding authentic, real-world applications or connections to the lab.

1.Safety and Equipment Report Form

Name _____

Campus _____   **Date** _____

A. Objective: To know where the safety features and some equipment is stored.

In the space provided below, draw your science lab room, labeling all the safety features listed, the **exit door(s), storage room door(s) if any, sink location and lab countertops (also called lab bench).** Locate and label the following: **fire extinguisher(s), first aid kit, fire blanket, fume hood, eye wash/shower, electronic balances, and chemical storage area.**

Complete the following information on the back of this page after you have completed your drawing.

B. Objective: To preview your lab manual and equipment we will be using. Using your lab manual and the classroom, identify the following equipment, *its use, and the lab(s) you will use it.* Additional equipment may be added to this, use the back of this page for those items.

Example Answer: Spectrometer, measures absorption of visible & UV light; Labs 10	
	List the content found in the appendix of this lab manual.
	Which of the community topics interest you in relation to chemistry and a creative or research project?

C. Measurement Report Form   Name_____

Your partner's name _____   Date_____

1. What were the objectives for the measurement lab?

2. Fill in the table with your measurements for the beakers' weight and fill in the data.

	Mass (g)	
Trial	Beaker 1	Beaker 2
1		
2		
3		
Average Mass		
Standard Deviation		

3. Fill in the table with your measurements of the graduated cylinder empty and filled with 10 mL of tap water.

Graduated Cylinder Mass (g)			
Empty	Filled 10 mL water	Mass of water only (g)	
			Average Mass of water (g)
			Standard Deviation of water mass

4. What was your calculated density of your water sample? (g/mL)

5. Which piece of glassware (beaker, flask, or graduated cylinder) was most accurate at measuring 50 mL?

6. Why was there a variation in the weight of your beaker each time you weighed it?

7. Were your measurements the same as others in the lab? Why?

8. What is the usual density of water given in science? (Look this up in your chemistry text, online, and be sure to cite where you found this information using these guidelines:

Book:	1. Author (year) *Title*. City: Publisher.
Research Article:	2. Author (year) "Article Title" *Journal title, volume,* pages.
Web page:	3. Author (year) "Page Title" Hosting institution, URL.

9. Is your calculated density the same value? Explain your answer, whether it is yes or no.

10. Explain what the terms "average" and "standard deviation" mean in relation to the data your own words.

2. Density Lab Report Form

I. Objectives

1. What were the Objectives of this Lab?

II. Safety Considerations *('as written' if not changes)*

III. Materials *('as written' if no changes)*

IV. Procedures *('as written' if no changes)*

2. Did you modify the procedures? If yes, explain here.

V. Conclusions

3. Record bulk density data in the following tables:

Bulk Density Data						
Name of sample: _____						
		Trial #			Average	Standard Deviation
		1	2	3		
Mass (grams)	Sample + Weigh Boat				-	-
	Weigh Boat or paper				-	-
	Difference					
	Bulk Volume (mL)					
	Bulk Density (g/mL)					

Bulk Density Data						
Name of sample: _____						
		Trial #			Average	Standard Deviation
		1	2	3		
Mass (grams)	Sample + Weigh Boat				-	-
	Weigh Boat or paper				-	-
	Difference					

	Bulk Volume (mL)					
	Bulk Density (g/mL)					

Bulk Density Data
Name of sample: _____

		Trial #			Average	Standard Deviation
		1	2	3		
Mass (grams)	**Sample + Weigh Boat**				-	-
	Weigh Boat or paper				-	-
	Difference					
	Bulk Volume (mL)					
	Bulk Density (g/mL)					

4. Explain the importance of knowing how to measure mass and volume correctly.
5. Record the mass density data in the following tables:

Mass (True seed) Density Data
Name of bean or seed Sample: _____

		Trial #			Average	Standard Deviation
		1	2	3		
Mass, g	**Sample + Weigh Boat**				-	-
	Weigh Boat or paper				-	-
	Difference					
Volume, mL	**Final**				-	-
	Initial				-	-
	Difference					
	Density (g/mL)					

Mass (True seed) Density Data
Name of bean or seed Sample: _____

		Trial #			Average	Standard Deviation
		1	2	3		
Mass, g	Sample + Weigh Boat				-	-
	Weigh Boat or paper				-	-
	Difference					
Volume, mL	Final				-	-
	Initial				-	-
	Difference					
	Density (g/mL)					

Mass (True seed) Density Data
Name of bean or seed Sample: _____

		Trial #			Average	Standard Deviation
		1	2	3		
Mass, g	Sample + Weigh Boat				-	-
	Weigh Boat or paper				-	-
	Difference					
Volume, mL	Final				-	-
	Initial				-	-
	Difference					
	Density (g/mL)					

6. Use density to explain why an ice cube made of tap water floats in tap water. Where could you apply your knowledge of materials densities in the real world in general or in your own life?

7. Summarize your results based on the lab objectives, your data, and an authentic application of the data *(5-10 sentences)*.

3. Chocolate Density! Report Form

I.   Objectives

    1.   What were the Objectives of this Lab?

    2.   Explain how density was affected with changing volume (and mass).

II.   Safety Considerations *('as written' if no changes)*

III.   Materials *('as written' if no changes)*

IV.   Procedures *('as written' if no changes)*

    3.   What is the importance of repeating a procedure several times in the lab?

V.   Conclusions

    4.   Fill in the Data table with your measurements.

	Mass (to the nearest 0.1 g)	**Length** (to the nearest 0.1 g)	**Width** (to the nearest 0.1 g)	**Height** (to the nearest 0.1 g)	**Volume** (to the nearest 0.1 cm³)	**Density** (to the nearest 0.1 g/cm³)
Whole Bar						
Half Bar						
Quarter Bar						
Twelfth Bar						

    5.   Compare the overall density of each of the different bar sizes.

	Whole Bar	Half Bar	1/4th Bar	1/12th Bar
Density				

    6.   Was the density significantly different (if the difference is less than 0.2 g/cm³, then we should assume that it's the same). Explain your answer

    7.   Summarize your results based on the lab objectives, your data, and an authentic application of the data *(5-10 sentences)*.

Name

Date

Class & Section

Your partner's name

4. Liquid Density Student Report Form

I. **Objectives**
1. What were the Objectives of this Lab?

II. **Safety Considerations** (*'as written' if not changes*)

III. **Materials** (*'as written' if no changes*)

IV. **Procedures** (*'as written' if no changes*)

V. **Conclusions**

Experiment A
1. Record your data in the data table.

	Beaker 1 Water		Beaker 2 Corn Syrup		Beaker 3 Vegetable Oil	
Small Object Tested	**Prediction**	**Actual**	**Prediction**	**Actual**	**Prediction**	**Actual**

2. Did you predict correctly? Yes, or no, explain your answer for each item.
3. Did your objects float in one liquid and not another? Why do you think this happened?
4. Explain using density of the liquids and the objects what you observed.

Experiment B

5. Record your data from the experiment here

Liquid	Predict which liquid is the most dense, least dense, and middle density	Actual density most dense, least dense, and middle density
Corn Syrup		
Vegetable Oil		
Water		

6. Did you predict correctly? Explain your answer.

Experiment C

7. What happened to the colored water? Did it stay in layers? Which layer was on the bottom? On the top? What does this tell you about the density of hot water compared to cold water?

8. What would happen if you left the cylinder out until the cold water warmed up and the hot water cooled off? Do more experimentation to find out!

Experiment D

9. Did adding salt and sugar to the water make the water denser or less dense? Explain your answer.

10. Which was denser, the salt water or the sugar water? Explain your answer.

Summary

11. Summarize your results based on the lab objectives, your data, and an authentic application of the data *(5-10 sentences)*.

5. Periodic Table of Videos Lab Student Report Form

I. **Objectives**

1. What were the Objectives of this Lab?

II. **Safety Considerations** *('as written' if not changes)*

III. **Materials** *('as written' if no changes)*

IV. **Procedures** *('as written' if no changes)*

V. **Conclusions**

A. **Questions to Answer for Elements 1 through 11**

1. What is produced when hydrogen is combusted?
2. What is produced when helium is combusted?
3. Lithium carbonate is used to treat manic depression. What is lithium's target in the brain?
4. What doesn't beryllium absorb?
5. Name one commercial product that contains boron.
6. Name one commercial product that contains carbon nanofibers.
7. How cold is liquid nitrogen in Fahrenheit, Celsius, and Kelvin?
8. What color is *liquid* oxygen?
9. What makes fluorine gas hazardous?
10. What color does neon emit when it is electrified?
11. Sodium is benign in one form but explosive in another. What are these two forms?

B. **Questions to Answer for the Other Three Elements**

For each of the other ***three elements*** that you chose to watch, provide:

12. The element name, symbol, and number
13. A *balanced* reaction involving that element
14. A brief description of the reaction
15. Give a source that describes the reaction. If you use your textbook, be sure to give the page number. If you use a website, you must give an accessible link to the site, as well as the date that you visited the site.

C. **Discussion:**

D. **References:**

6: Qualitative Testing of Water Student Report Form

I.   Objectives

    1.   What were the Objectives of this Lab?

II.   Safety Considerations *('as written' if not changes)*

III.   Materials *('as written' if no changes)*

IV.   Procedures *('as written' if no changes)*

V.   Conclusions

    2.   Fill in the data table.

Water Sample Name	Location	
Temperature of water (in field)		
Ion/Chemical Test	**Color Observed**	**Present (+) or absent (-)**
pH		
Conductivity		
Ammonia		
Chlorine		
Iron		
Calcium		
Nitrate		
Phosphate		
Sulfide		
Chromate		
Copper		
Cyanide		
Pesticide		

    3.   What ions were present in your water sample?

    4.   Regardless of a positive or negative test, how would nitrates show up from in your water samples?

    5.   Regardless of a positive or negative test, how would pesticides show up in your water samples?

    6.   What makes these tests qualitative instead of quantitative test?

    7.   Why do some ionic compounds dissolve in water while others remain solid or form precipitates?

8. Were any microorganisms present in your water sample? If yes, identify on which petrifilm you had growth, and take a picture of the surface of the petrifilm and insert here, labeling appropriately. *(Alternatively, sketch a picture of the surface.)*
9. Was any of the test results a surprise to you? Explain your answer.
10. What do you think about the quality of the drinking water at your home? In your town? Here at the school? Explain your answers.
    11. Summarize your results based on the lab objectives, your data, and an authentic application of the data *(5-10 sentences)*.

Name

Date

Class & Section

Your partner's name

7. Water Purification Student Report Form

I. **Objectives**

   1. What were the Objectives of this Lab?

II. **Safety Considerations** *('as written' if not changes)*

III. **Materials** *('as written' if no changes)*

IV. **Procedures** *('as written' if no changes)*

V. **Conclusions**

   2. Present your data filtration/purification data in table form.

   3. If you screened for microorganisms, present your results here, inserting photograph or sketch as well.

   4. Were you able to clean your foul water sample? Is it possible there are still contaminants in your filtered water? How could those unseen contaminants be removed from the water? Explain your answer with evidence from your data.

   5. Where does your drinking water come from? How is it treated before it is available for drinking? Compare your filtration set-up to a water treatment plant. If you have not visited a water treatment plant, use internet resources for the comparison. You may use drawings to enhance your comparison. Be sure to include the reference(s) for your information.

   6. List five (5) chemicals and or biological contaminants that is tested for and reported on in the Environmental Protection Agency's National Primary Drinking Water Regulations: Federal water quality standards found at https://www.epa.gov/ground-water-and-drinking-water/national-primary-drinking-water-regulations and locally in your community.

   7. When you washed the filter with distilled water, was the water clean after passing through the filter (the filtrate)? Explain your answer.

   8. What water topics are you familiar with in your community? Region? State or Nationally? Which one community water topic personally interests you and why?

   9. Summarize your results based on the lab objectives, your data, and an authentic application of the data *(5-10 sentences)*.

8. Qualitative Soil Testing Student Report Form

I. **Objectives**

    1. What were the Objectives of this Lab?

II. **Safety Considerations** *('as written' if not changes)*

III. **Materials** *('as written' if no changes)*

IV. **Procedures** *('as written' if no changes)*

V. **Conclusions**

    2. Report your results in a table.

    3. What role does pH play in plant growth?

    4. Are there specific amounts of phosphorus, nitrogen, and potassium that will maximize plant yield? Explain your answer.

    5. What can be done to change the pH of soil that is too acidic? Too basic?

    6. If a soil is deficit in (a) nitrogen, (b) phosphorus, and (c) potassium, what can be done to increase quantities?

    7. What commercial crops are grown locally? How can the information and methodology learned in this lab help agriculture in your community?

    8. Do backyard gardeners need to check their soil nutrients? Explain your answer.

    9. What vegetables and fruits do local gardeners grow for their own personally or farmer's market use in your area? *(As opposed to the large commercial crops.)* How can small farmers use the information from this lab?

    10. Regardless whether you found pesticides present in your sample or not, is the pesticide a contaminant if found in the soil?

    11. Are the pesticides we tested for toxic if present in the soil to the plants?

    12. Are they pesticides we tested toxic to humans?

    13. Were any microorganisms present in your soil sample? If yes, identify on which petrifilm you had growth, and take a picture of the surface of the petrifilm and insert here, labeling appropriately.

    14. Summarize your results based on the lab objectives, your data, and an authentic application of the data *(5-10 sentences)*.

9. Herbicide Bioassay Student Report Form

I. **Objectives**

1. State and explain how you met the objectives of this lab.

II. **Safety Considerations** (*'as written' if not changes*)

III. **Materials** (*'as written' if no changes*)

IV. **Procedures** (*'as written' if no changes*)

V. **Conclusions**

2. Report your data and observations:

a. Control seedlings Data

	Number of Seedlings that Sprouted	Average Length of Roots	Other Observations
Distilled water			

Drawings:

b. NaCl seedlings Data:

Final [NaCl] (mM)	Number of Seedlings that Sprouted	Average Length of Roots	Other Observations
0.0			
25			
50			
75			
100			
250			

Drawings (NaCl):

c Herbicide Seedlings Date:

Final [Herbicide] (mg/L)	Number of Seedlings that Sprouted	Average Length of Roots	Other Observations
0.0			
0.00010			
0.0010			
0.010			
0.10			
1.0			

Drawings:

3. What is the IC50 for NaCl? Compare "sprouting" with "root length."
4. What is the IC50 for your herbicide? Compare "sprouting" with "root length."

5. Include your drawings of your control and partially inhibited seedlings.
6. Compare your IC50s with the corresponding animal LD50s and comment on the difference.
7. How would you improve the assay if you were to do it again?
8. How could/does this lab, using herbicides, relate to your daily life? Write a short paragraph (3-7 sentences) connecting the topic of this lab to something in your community.
9. Summarize your results based on the lab objectives, your data, and an authentic application of the data *(5-10 sentences)*.

10. Chromatography & Spectrometry Student Report Form

I. Objectives

1. State the objectives and explain how you met the objectives of this lab.
2. Do absorption spectra identify specific chemicals? Explain your answer.

II. Safety Considerations *('as written' if not changes)*

III. Materials *('as written' if no changes)*

IV. Procedures *('as written' if no changes)*

V. Conclusions

3. Draw your chromatography results (or attach a photograph).
4. How many different pigments separated on your paper chromatogram? Were you able to identify the different pigments with their Rf values? Explain your answer, using your Rf values data.
5. Attach your absorption spectra graphs with the maximum wavelength (s) identified.
6. Compare the results from the different dilution samples. Were the spectra similar or different as you diluted the extract?
7. As you increased the extract portion of the sample, what happened to the absorption value?
8. Describe your own experience using native dyes.
9. Do you know of other plants or materials than what we used in the lab that are used for dying fabrics or added to food for color? Please share those with us.
10. Summarize your results based on the lab objectives, your data, and an authentic application of the data *(5-10 sentences)*.

11. Chemical Reactions Student Report Form

I. **Objectives**
   1. What did you learn besides the stated objectives in the lab?

II. **Safety Considerations** *('as written' if not changes)*

III. **Materials** *('as written' if no changes)*

IV. **Procedures** *('as written' if no changes)*

V. **Conclusions**
   2. Show your data in the following tables. If you did not use all of the chemicals just draw a X through the extra tables or if turning in digitally delete.

Amount of NH_4NO_3 or NH_4Cl

Temperature, ° C	Beaker$_A$,	Beaker$_B$,	Beaker$_C$,	Beaker$_D$,
Initial				
Final *(Lowest Temperature Measure)*				

Amount of NH_4NO_3 or NH_4Cl

Temperature, ° C	Beaker$_A$, 50 g	Beaker$_B$, 100 g	Beaker$_C$, 125 g	Beaker$_D$, 150 g
Initial				
Final *(Lowest Temperature Measure)*				

Amount of Na_2CO_3 or $CaCl_2$

Temperature, ° C	Beaker$_A$	Beaker$_B$	Beaker$_C$	Beaker$_D$
Initial				
Final *(Lowest Temperature Measure)*				

Temperature, ° C	Beaker$_A$	Beaker$_B$	Beaker$_C$	Beaker$_D$
Amount of Na$_2$CO$_3$ or CaCl$_2$				
Initial				
Final *(Lowest Temperature Measure)*				

3. How did you know a chemical event had taken place?
4. Which, if any, amount of NH$_4$NO$_3$ (or NH$_4$Cl) chemicals produced the coldest temperature with the water? Explain why we kept the water volume constant. *If* you repeated the cold pack reactions with the other chemical, which of the two chemicals produced the coldest temperatures?
5. What is ammonium nitrate, NH$_4$NO$_3$, used for in real world?
6. What is ammonium chloride, NH$_4$Cl, used for in the real world?
7. Which, if any, amount of Na$_2$CO$_3$ (or CaCl$_2$) chemicals produced the warmest temperature with the water? Explain why we kept the water volume constant. *If* you repeated the hot pack reactions with the other chemical, which of the two different chemicals produced the warmest temperatures?
8. Sodium carbonate, Na$_2$CO$_3$, can be purchased as washing soda. What purpose does it serve in your laundry? Are there any other uses for Na$_2$CO$_3$?
9. What is a real world use of calcium chloride, CaCl$_2$?
10. Summarize your results based on the lab objectives, your data, and an authentic application of the data *(5-10 sentences)*.

Student Lab Template

Your Name
Date
Class & Section
Partners Name(s)

Title

Objectives *Given in lab*

Safety Considerations, *if any changes otherwise, 'as written'*

Materials *if any changes otherwise, 'as written'*

Procedures *if any changes otherwise, 'as written'*

Data, *organized, preferably in a table or graph*

Conclusions *answers to given questions and a summary*

References, *if used any outside sources or obtained data from a partner or another group*

Appendix E. PHOTO CREDITS

Photographs on pages 6, 19, 20, 38, 60, 70, 169, and 176 were taken by Beverly DeVore-Wedding.

The photograph on page 68 was taken by Mark Griep.

The photograph of the spectrophotometer on page 176 was taken by Janyce Woodard.

Appendix F. HISTORY AND BACKGROUND OF THE EXPERIMENTS

The experiments in this lab manual are a compilation of the authors' labs and interests. Our primary goal was to incorporate cultural and community interests into standard laboratory experiences. The experiments in this lab manual were compiled from several sources, which we then modified by adding community connections and adapting them based on feedback from the instructors and students at Little Priest Tribal College (LPTC) and Nebraska Indian Community College (NICC). The main sources and inspirations for the labs were: (1) Dr. Mark Griep's Chem 105 Lab Manual at the University of Nebraska-Lincoln; (2) Ms. Janyce Woodard's organic labs from her organic and biochemistry course at Little Priest Tribal College; and (3) the RISE Experiments from Turtle Mountain Community College.

The introduction referring to working in a laboratory classroom, preparing for labs, and participating safely are adapted from Dr. Griep's lab manuals.

Experiment 1, Safety, Equipment, and Measurement, was written to introduce students to laboratory safety procedures and equipment. The safety part of the lab is modified from Dr. Griep's lab manuals second edition published in 2014 and Ms. Woodard's safety introduction at LPTC. The lab room drawing with safety features is adapted from Dr. DeVore-Wedding's teaching materials. The measurement portion of this lab comes from both Woodard's and DeVore-Wedding's lab instruction.

Experiment 2, Density, is adapted from Griep's lab manual with an expansion of the community connection.

Experiment 3, Chocolate Density, and Experiment 4, Liquid Density, were developed by DeVore-Wedding to reinforce the concept of density. The ideas came from a Google search of density labs. The labs were written to incorporate student interest, measurement and calculation practice, and to address several of the community connections.

Experiment 5, Periodic Table Video, was created by Griep and Scott Raber. To learn more about these award-winning video series, read Haran and Poliakoff (2011) *Science* 332, 1046-1047.

Experiment 6, Water Quality Analysis, was developed by DeVore-Wedding and Woodard in 2015 to meet the community connection topic of water sources and introduce students to common ions and qualitative tests while introducing water chemistry.

Experiment 7, Water Purification, was first a student project, then developed as a connection to the community topic of water sources. The lab was modified from the American Chemical Society's "Chemistry in the Community's" (1993, 2nd edition) lab *"Foul Water."*

Experiment 8, Soil Quality Analysis, originally started as Griep's "Nitrate Content in Soil" lab and was modified to include testing for phosphorus and potassium as well.

Experiment 9, Herbicide Bioassay, was adapted from Cornell University's "Environmental Inquiry" website at http://ei.cornell.edu/toxicology/bioassays/lettuce/RefTest.asp and modified by adding the community connection and a screening for microorganisms.

Experiment 10, Plant Pigments: Extraction, Chromatography, & Spectrometry, was adapted from Griep's "Spectroscopy of Natural Dyes in Fruit Drinks" to include native plant pigments and chromatography to increase student laboratory experiences and connect with community topics. Ideas for adaptation also came from a lab in "RISE Experiments, Introductory Organic and Biochemistry" (Turtle Mountain Community College, 2006), which were developed by North Dakota Tribal College students and faculty.

Experiment 11, Endothermic and Exothermic Reactions –Hot & Cold Packs, was developed from a demonstration "CHEM-PACS *Practical Activities with Common Substances*" (Berger, Bryce, Gettman, O'Brian, and Squires, 1989) from Flinn Scientific. Students were intrigued so it was revised to introduce limiting reagents.

Experiment 12, Molar Mass of Butane Lighters, was adapted from the UNL Lab Manual.

Experiment 13 and 26, First Semester Creative Project and Presentation, was from Woodard's class at Little Priest Tribal College as a final project for labs. Information about making a poster is from Griep's REU projects.

Experiment 14, Acid & Base Indicators, was adapted from a lab in "RISE Experiments, Introductory Organic and Biochemistry" (Turtle Mountain Community College, 2006).

Experiment 15, Ascorbic Content in Traditional Native Foods, was adapted from a lab in "RISE Experiments, Introductory Organic and Biochemistry" (Turtle Mountain Community College, 2006), using common lab protocols using starch and an iodine solution.

Experiment 16, Qualitative Tests for Alcohols, was adapted from a lab in a lab in "RISE Experiments, Introductory Organic and Biochemistry" (Turtle Mountain Community College, 2006), connecting to our community topics.

Experiment 17, Qualitative Tests for Aldehydes and Ketones, Woodard adapted from E. Boschmann & Norman Wells (1990), *Chemistry in Action: A Laboratory Manual for General, Organic, and Biological Chemistry*, 4th Edition, page 309.

Experiment 18, The Effect of Alcohol on Betacyanin Extraction, was adapted from a lab in "RISE Experiments, Introductory Organic and Biochemistry" (Turtle Mountain Community College, 2006).

Experiment 19, Preparation and Identification of Esters, was adapted from "RISE Experiments, Introductory Organic and Biochemistry" (Turtle Mountain Community College, 2006).

Experiment 20, Synthesis of Aspirin and Wintergreen Oil, was adapted from the UNL Chem 110 Lab manual and "RISE Experiments, Introductory Organic and Biochemistry" (Turtle Mountain Community College, 2006).

Experiment 21, The Science of Soap Making, was adapted from the UNL Chem 110 Lab manual and "RISE Experiments, Introductory Organic and Biochemistry" (Turtle Mountain Community College, 2006).

Experiment 22, Transesterification of Oils in Native Plants, was adapted from "RISE Experiments, Introductory Organic and Biochemistry" (Turtle Mountain Community College, 2006).

Experiment 23, Extraction of Caffeine from Yerba Mate, was adapted from the UNL Chem 110 Lab manual and "RISE Experiments, Introductory Organic and Biochemistry" (Turtle Mountain Community College, 2006).

Experiment 24, The Effect of Heat on Enzymes Found in Native American Fruits, was adapted from "RISE Experiments, Introductory Organic and Biochemistry" (Turtle Mountain Community College, 2006).

Experiment 25, DNA Extraction, was a student project based on Internet resources. We modified using ideas from "RISE Experiments, Introductory Organic and Biochemistry" (Turtle Mountain Community College, 2006) and added in spectrometric analysis using two resources: Chem. 247.53 Biophysical Chemistry. (2010). *Isolation, spectrophotometric Analysis and Melting of Onion DNA* (https://chem247.files.wordpress.com/2007/09/chem-247-dna-lab.pdf) and Promega corporation. (2018). *How do I determine the concentration, yield and purity of a DNA sample?* (https://www.promega.com/resources/pubhub/enotes/how-do-i-determine-the-concentration-yield-and-purity-of-a-dna-sample).

www.ingramcontent.com/pod-product-compliance
Lightning Source LLC
Chambersburg PA
CBHW081556220526
45468CB00010B/2671